D0991952

Creating the American Junkie

Creating the American Junkie

Addiction Research in the Classic Era of Narcotic Control

Caroline Jean Acker

THE JOHNS HOPKINS UNIVERSITY PRESS
Baltimore & London

Printed in the United States of America on acid-free paper
9 8 7 6 5 4 3 2 1

The Johns Hopkins University Press
2715 North Charles Street
Baltimore, Maryland 21218-4363
www.press.jhu.edu

Library of Congress Cataloging-in-Publication Data

Acker, Caroline Jean, 1947–
Creating the American junkie : addiction research in the classic era of narcotic control / Caroline Jean Acker.
p. ; cm.
Includes bibliographical references and index.
ISBN 0-8018-6798-3 (hardcover : alk. paper)
1. Heroin habit—Social aspects—United States. 2. Heroin habit—Research—United States. 3. Narcotic habit—Research—United States. 4. Narcotic addicts—Research—United States. 5. Narcotic addicts—Government policy—United States.
[DNLM: 1. Heroin Dependence—history—United States. 2. Drug and Narcotic Control—history—United States. 3. Heroin Dependence—etiology—United States. 4. Research—history—United States. WM 11 AA1 A182c 2001] I. Title.
HV5822.H4 A375 2002
362.29'3'0973—dc21

2002001520

A catalog record for this book is available from the British Library.

In memory of

David Gordon Acker

Greta Gordon Acker

Jack David Pressman

and

Glenn Albinger

Contents

Creating the American Junkie

Introduction

The American junkie is a product of American history. The heroin addict—typically portrayed in movies, newspapers, and folklore as a heroin-addicted male urban hustler—emerged during a period when the marketing of opiates and the management of urban vice was undergoing profound transformations. These changes created the context for a particular pattern of exposure to now criminalized opiates and of managing the highly stigmatized addiction that resulted from that exposure. The result was a cohort of morphine and heroin addicts, mostly male, concentrated in urban centers, who used hustle and street smarts to navigate an underground market for the drug they had become dependent on. The oldest of these addicts began using opiates and cocaine before these drugs became illegal; indeed, the passage of laws against their nonmedical sale and possession was in part a reaction to them. But as social disapproval hardened into criminal sanctions, and as the move to prohibit opiates drove the market for them further underground, addicts developed the skills to maneuver in a tricky world of passwords, shifting sales locations, and undercover narcotics officers in order to sustain their drug habits. A demographic subgroup reacting to a particular set of public policy and historical changes, these addicts gave rise to an image of deviance that has shaped American drug policy ever since and helped reinforce the moral underpinnings of the war on drugs. If heroin addicts were really as deviant and unreclaimable as this image held, then a war on drugs that resulted in their demonization and incarceration might seem justified. The truth is more complex, interesting, and troubling than this simple morality tale implies.

In 1890, opiates were sold in an unregulated medical marketplace. Physicians prescribed them for a wide range of indications, and pharmacists sold them to individuals medicating themselves for physical and mental discomforts. These patients and customers were most typically middle-class, middle-aged women taking morphine to relieve the pain of menstrual cramps or assuage domestic or social anxieties. By 1910, this pattern was in decline, as concerns about iatrogenic addiction (addiction caused in the course of medical treatment) led physicians to reduce the prescribing of opiates. Over the next few decades, reforms enacted by physicians and pharmacists, as well as legal restrictions on the sale of opiates, caused the medical market for opiates to shrink, with a sharp reduction in the number of people who became addicts as a result of medical use of opiates.

In the same period, the sale of opiates outside of medical channels increased, especially in northeastern American cities. Recreational use of opiates shifted from the smoking of opium centered in urban Chinatowns to nonmedical consumption of morphine, swallowed or injected subcutaneously (under the skin rather than into a vein). When heroin was marketed as a cough remedy, beginning in 1898, its euphoric effects attracted young working-class men from crowded urban neighborhoods seeking amusement. By 1910, heroin sniffing had become a well-established component of a repertoire of drug-use patterns attached to a shifting urban entertainment scene that included dance halls, poolrooms, and vaudeville theater. Sniffers often became injectors, and in the 1920s, intravenous, rather than subcutaneous, injection became increasingly common. As the older cohort of individuals addicted to opiates through medical use died off and fewer new ones replaced them, the most typical opiate addict became a working-class male in young adulthood living in a working-class neighborhood whose amusement venues included a variety of morally suspect choices by the standards of the Protestant middle class: pool halls, bars, brothels, restaurants, theaters, dance halls, and perhaps a racetrack.

Nevertheless, the distinction between medical and recreational addicts was not actually as sharp as this brief account suggests. Some people who began taking opiates in medical settings moved into the social world connected to the emerging illicit market; and most addicts who began using opiates purely for pleasure found themselves burdened with a habit that took on the characteristics of chronic illness. In addition, some who first sampled opiates in a purely recreational setting found them compelling not because of euphoric effect but because they instantly relieved long-standing depression or anxiety. However, the perception that the newer urban male addicts were fundamentally different from individuals who had become addicted medically would come to dominate public and academic views.

To many upper-middle-class Anglo-American Protestants, the traditional elites in American cities, this pattern of drug use reflected an alarming increase in vice, a problem that added to their anxieties about the profound social transformations resulting from industrialization, urbanization, and new patterns of immigration. Their concerns resulted in an array of Progressive Era reform movements. A move against prostitution was for many a central plank in a broad range of efforts to quash urban vice. Just as reformers had urged the building of water filtration and sewer systems to help protect the city against diseases associated with overcrowding, so too Progressives urged new laws to lance the boil of urban vice. Around 1910, as reformers sought to suppress prostitution altogether, American cities moved against the tolerated, segregated vice districts where brothels were located. These efforts occurred simultaneously with campaigns to prohibit nonmedical use of alcohol, opiates, and cocaine, as well as agitation against tobacco, gambling, gangs, risqué entertainment, and new forms of dancing. America's national experiment with alcohol prohibition and the more enduring criminalization of the nonmedical use of opiates and cocaine were among Progressive Era reform achievements.

These reforms created upheaval in Tenderloin districts in many American cities as reformers struggled with urban political machines whose support derived from both working-class immigrants and vice entrepreneurs, many of whom were denied entry into more respectable job markets, to enact stable regimes of vice suppression. Indeed, the frequent willingness of local governments to act vigorously against their constituents led Progressive reformers to urge such federal legislation as the Mann Act, which prohibited transporting women across state lines for immoral purposes, and alcohol prohibition. In the two decades following World War I, Progressive Era reforms resulted in a new pattern in the management of urban vice in American cities, in which some norms were liberalized while other behaviors were more harshly condemned. The cases of alcohol and tobacco offer instructive parallels to that of cocaine and opiates. Prohibition of alcohol was repealed in 1933, and tobacco had never aroused enough reformist ire to culminate in widespread prohibition. Use of opiates and cocaine, on the other hand, remained criminalized and underground, where it increasingly became a marker of profound deviance. Analogously, in the 1920s, sexual mores loosened, so that a broader range of heterosexual and social behaviors were consistent with respectability than before World War I; but prostitution, or outright sale of sex, was forced from unofficial tolerance into criminal management. The result was to destroy such social support as prostitutes sustained in madam-managed, geographically stable brothels; thereafter, many prostitutes worked in more dangerous and socially isolated situations, with pimps as managers. Homosexuality followed a somewhat different trajectory toward a similar result. Grouped among other objects of vice reform in the Progressive Era, male homosexual behavior (interpreted as effeminacy in men) was subject to vice reformers' surveillance and intermittent police action. In the 1920s, prominence of female impersonators' acts in the Prohibition nightclub scene, culminating in a "pansy craze" around 1930, suggested the liberalizing influence might include the gay world. However, following repeal of alcohol prohibition, new regulatory oversight of bars drove "disorderly conduct," including gay behavior, out of night spots formerly patronized by both gays and straights. The result was the emergence of gay-only bars, which had to shift locations in response to police raids and closures. Gays, too, developed networks of communication to stay apprised of the resulting changes of venue. Following World War II, homosexuality, like drug addiction, symbolized profound, irredeemable deviance.[1]

Thus, the American junkie arose in an urban vice subculture more harshly repressed than it had been just a generation earlier. To sustain an addiction to opiates in such an environment required a set of skills that made for success in the constantly shifting scene of America's urban vice districts. The imperatives of heroin addiction—obtaining steady supplies of illegal drugs, which became ever more expensive as a result of public policy, acquiring the means to pay for them, securing private locations to inject them, hiding addiction from family and co-workers, dodging undercover police and sizing up dealers, and maintaining social ties with fellow heroin users—intensified the value of street smarts for the addict. Easy sociability, psychological intuitiveness, and an ability to maintain the cover stories and identities called for in different settings all increased the addict's chances for maintaining his or her habit and avoiding such negative consequences as arrest, job loss, and family ostracism. Some addicts managed to live in this way for long periods, but the tightening demands of progressive addiction drove many further outside the bounds of respectability and legality, and their behavior strengthened social disapproval of addiction itself.

Urban vice reformers also created a new context for the academic and clinical study of opiate addiction. The Progressive Era and the period following World War I witnessed ferment and reform in the medical and scientific professions. This process affected anti-vice campaigns as social reform increasingly became an arena of professional rather than private voluntary activity. Key to this shift was the rising power in the same period of scientific medicine and of science more generally. Researchers in a number of disciplines provided explanations of and suggested solutions to a wide range of social problems, from the control of venereal disease to salubrious tenement design. As a problem involving substances used as medicines and affecting both body and mind, opiate addiction drew the attention of researchers and clinicians interested in both the physiological and the behavioral effects of drugs. Scientists and clinicians who focused their disciplinary lenses on opiate addicts in this period concluded that characteristics of this cohort held true of opiate addiction more generally.

Researchers' structural and personal links to emerging federal drug policy agencies formed a conduit whereby research results strengthened moral disdain and harsh policies toward opiate addicts. In the interwar period, the impact of research on policy was mediated primarily in the arena of public health and secondarily in that of drug law enforcement, and the presence of both of these components of federal policy exemplified an enduring tension between medical and criminal justice management of addiction.

This book analyzes the relationship between the first cohort of illicit opiate users in the United States and the researchers—pharmacologists, psychiatrists, and sociologists—who studied them. Examining the connections between these researchers and federal policy makers, it also shows how research findings in psychiatry and pharmacology bolstered a stiffening federal policy of punishing opiate addicts that, ironically, helped create a more tightly knit, if socially disconnected, junkie subculture. In contrast, sociological studies of opiate addicts provided the basis for a critique of federal policy.

The junkie of the post–World War I period, then, was not the inevitable result of a particular character or personality type, or of inherent criminality. Nor was his or her addiction simply the outcome of the pharmacological effects of heroin. Rather, the junkie deployed a set of coping strategies for managing an opiate addiction that had emerged in a particular social and historical context. In other words, the meaning of drug use depends crucially on the social and cultural settings of that use.[2] At the macro level, market structures for drugs influence who will use them and in what ways. Thus, changes in public policy and in medical practice transformed the market in opiates between 1890 and 1930; these changes shaped the urban social milieu in which heroin trade was concentrated and influenced how heroin addiction was culturally defined throughout the rest of the twentieth century.

At the micro level, an individual's immediate social environment and his or her roles within it provide a context for understanding the meanings invested in the use of psychoactive drugs. Thus, both the anxious social matron of the late nineteenth century and the socially marginal laborer of the early twentieth might find opiates compelling because of their powerful ability to soothe and calm, to create a sense that all is right with the world. But the differences in individual identity and social context made for profound differences in these two individuals' experience of drug use. For the matron, subcutaneous injections of morphine made it easier to live up to the social expectations she felt were placed on her. If her habit became known to others besides her physician, she typically became an object of pity. In contrast, heroin likely made it easier for a twentieth-century housepainter, say, who gambled and enjoyed burlesque shows, to reject mainstream society's expectations of him. If caught, however, he faced not only legal sanctions but also profound stigma. The nineteenth-century matron obtained inexpensive opiates at little social cost; the twentieth-century urban laborer was forced to secure opiates that were increasingly costly in both financial and social terms.

One hundred years ago, the image of the junkie had not entered the popular culture; today, the junkie is a deeply familiar symbol of irredeemable deviance, of a life turned upside down as the seeking and taking of drugs overwhelm love, family, and work. This image of addiction has broadened beyond the heroin user; users of other illicit psychoactive drugs, most notoriously in recent years, crack cocaine, are now included in it. Whatever the drug or the label attached to the user, the stereotype is of an individual taken over by drugs who becomes oblivious to expected social roles and the normal demands of life.

Drugs, licit and illicit, have been the objects of moral panics in the United States since the rise of the temperance movement in the nineteenth century, but most especially since the Progressive Era. In recent decades, one drug has succeeded another in the headlines, each accompanied by admonitions that it is uniquely addictive or likely to produce bizarre behavior. Media coverage of angel dust (phencyclidine, or PCP) in the late 1970s stressed the latter, with stories of prisoners snapping their wrists out of handcuffs.[3] The dominant theme with crack cocaine in the 1980s and 1990s was its quick and vicelike grip on the addict, elaborated with stories of women, usually black, who compulsively traded sex for crack and who bore infants addicted and damaged in utero because of the mother's drug use during pregnancy. As crack use diminished in the late 1990s, stories of "ice," methamphetamine, picked up the theme of rapid addictiveness and, instead of playing up the exoticness of crack houses in urban ghettoes, pointed to danger in the heartland when clandestine methamphetamine laboratories were exposed in midwestern towns.

All of these are variants on the junkie image that emerged in the first decades of the twentieth century, when opiate addiction ceased to be framed as a morally enslaving condition that trapped the unwary, but pitiable, and became a mark of depravity and degradation. That transition set the stage for what David Courtwright has called the classic era of narcotic control, the period from 1919, when a pair of Supreme Court decisions prohibited addiction maintenance as a treatment method, until about 1964, when methadone maintenance treatment was introduced.[4] In those four and a half decades, federal drug policy centered on rigid enforcement of laws criminalizing sale, possession, or use of opiates unless medically authorized. Treatment options all but dried up, and addicts navigated a precarious world of illicit drug sales, clandestine networks of fellow users, and the likelihood of incarceration, whether in a local lockup or in one of the Public Health Service's narcotics hospitals.

The durability of drug use as an icon of hopelessness has several implications. The addict has been a potent negative symbol of inverted social roles as addicts abandon responsibilities and sever social and affectional ties. Compulsive drug use has functioned as a metaphor of consumption out of control in a society whose culture is increasingly dominated by consumerism.[5] Social utility aside, the image of the junkie draws some of its authority from its partial grounding in medicine and science. A durable source for this image of the junkie lay in the disciplinary aspirations of psychiatrists, pharmacologists, and, to a lesser degree, sociologists in the first decades of the twentieth century. This was a time of forceful discipline building and reform in medicine and the natural and social sciences. The professional image of the heroin addict has lacked some of the lurid features of the popular version, but it has shared both the idea that the addict's behavior represents an inversion of normal and expected social roles and deep pessimism regarding the addict's prospects for recovery. The legacy of this profoundly negative professional and policy construction of the addict in the interwar years lasted well beyond World War II in lengthy incarceration of addicts, and in physicians' reluctance to treat addicts as patients. Furthermore, medical researchers virtually abandoned the problems of opiate addiction outside of the Public Health Service's Addiction Research Center, a site located until the mid 1970s at an institution closely linked to the federal government's vigorous enforcement of the prohibition on opiates, the Public Health Service Narcotic Hospital in

Lexington, Kentucky. The professional construction of the addict emerged from the encounter between clinicians and researchers, on the one hand, and the cohort of opiate addicts who first faced the challenges of habit management under a regime of drug prohibition, on the other.

This is not to say, as some liberal policy analysts argued in the 1970s, that the tough young male from a deteriorated inner city neighborhood who typified the heroin addict of the 1920s through the 1950s was simply a perverse consequence of the prohibition of nonmedical use of opiates and cocaine by the Harrison Narcotic Act of 1914.[6] This study adds to a more recent body of scholarship showing that a cohort of young men (and to a lesser degree, young women) began using opiates and cocaine for pleasure from the 1890s on in the dance halls, pool halls, and other entertainment venues springing up in working-class neighborhoods in American cities in the Progressive Era.[7] This first cohort of what we now call recreational drug users included many who engaged in the broad range of vices that troubled elite middle-class groups: from unescorted mingling of the sexes in dance halls through smoking cigarettes, gambling, drinking alcohol, engaging in casual sex (some commercial, some not) to overtly criminal activities. The passage of state and then federal laws against the nonmedical use of opiates and cocaine was in part a response to this pattern. More important, at the same time, banning legal access to the drugs drove addicted users to the emerging illicit market, which also supplied occasional users and the social networks in which first-time use typically occurred; thus, prohibition and the illicit market intensified trends that predated passage of the Harrison Act.

The Progressive Era rise in opiate and cocaine use was the first of two major demographic shifts in the use of the illicit psychoactive drugs in the United States in the twentieth century. This wave began before and influenced the passage of the Progressive Era laws that made these drugs illicit; it coincided from its earliest years with a decline in the overall prevalence of opiate addiction as widespread opiate use in medical or quasi-medical contexts waned.[8] Numbers of this newer type of recreational user declined somewhat in the 1930s and more sharply in the 1940s, when World War II disrupted international trafficking patterns.[9] As trafficking networks were reestablished after the war, opiate use began to rise again. It was still largely—but by no means exclusively—confined to deteriorated urban neighborhoods, which by now were increasingly inhabited by African Americans.[10] Throughout this "classic era," criminal justice approaches to opiate addiction virtually eclipsed medical ones.

The second major shift in American drug use patterns occurred as an explosion of drug use by middle-class youth in the 1960s and 1970s. As drug use spilled out of isolated and marginalized groups into more mainstream sectors of American society, addiction was redefined in less stigmatizing terms. A broadening menu of available psychoactives and a growing trend of involvement with more than one drug contributed to the development of a generic concept of dependence that could embrace not only a wide variety of drugs, but also compulsive behaviors such as gambling.[11] The introduction of methadone maintenance treatment in 1964 signaled the end of Courtwright's classic era and the beginning of an expansion of treatment methods and access to treatment. Richard Nixon amended the decades-long federal reliance on supply-side controls by providing federal resources for community-based treatment in addition to strengthening enforcement. A grassroots movement of free clinics, most famously the Haight Ashbury Free Clinic in San Francisco, began in the late 1960s to meet the medical needs of young people who had cut themselves adrift from families and mainstream institutions; in this clinical setting, a new set of encounters between clinicians and drug users produced a new definition of addiction.[12]

In the 1970s and 1980s, demand for drug treatment from individuals with jobs and health insurance coincided with hospitals' need to develop profit sectors in the face of rising health care costs; the result was a profusion of private sector drug treatment centers that offered detoxification, counseling, and other treatment services. These are currently undergoing curtailment with the spread of managed care, but even at their peak, they coexisted with continuing and pervasive stigmatization of drug users both as citizens and as patients. We now have in the United States a two-tier system of response to drug dependence: treatment for the middle and upper classes and incarceration for most others, including the poor, the uninsured, ethnic minorities, and immigrants. Employment status, race, gender, and class all influence which response an individual encounters.[13]

It is the argument of this book that the interaction between the first cohort of modern recreational opiate users and researchers in psychiatry and pharmacology produced a construction of the tough urban addict, not as straightforward description of psychological or physiological reality, but as a complex amalgam of observations and disciplinary concerns, which, in turn, were closely linked to the creation of a federal treatment and enforcement structure grounded in a supply-side policy approach to dealing with unauthorized use of opiates and cocaine. My aim is neither to discount the difficult behavior of many addicts as they sought treatment for their habits, nor to deny the difficulties faced by conscientious clinicians and researchers struggling to deal with a poorly understood condition that eluded their best efforts at understanding and alleviation. Rather, I argue that the construction of addiction as a physiological and psychological phenomenon in the 1920s and 1930s was profoundly influenced by (1) the professional concerns of psychiatrists seeking to modernize a medical specialty seen as lagging behind the scientific transformation of medicine; (2) the disciplinary ambitions of pharmacologists to build a drug-development infrastructure in the United States to rival that of Germany; and (3) a campaign by the American Medical Association to end the pattern of iatrogenic addiction that had characterized the late nineteenth century by reducing the prescription of opiates and absolving physicians in private practice from the necessity of treating troublesome addicts as patients. All three of these interests proved consistent with the criminalization of nonmedical opiate use and with specific federal policies, including the Treasury Department's Division of Narcotics' policy of ending ambulatory treatment of addicts and the creation by the Public Health Service (then also in the Treasury Department) of two institutions of incarceration for addicts, which functioned as both prisons and hospitals.

Drug use as a social behavior became a problem of interest to a small group of sociologists in the 1930s. Their work provided the basis for a critique of federal policy; but it emerged in a period when an ascendant sociological functionalism implicitly aligned itself with prevailing policy as it prioritized social harmony above diversity and cast many dissenting behaviors as deviant.

A specific set of ideas about addiction emerged from and reflected the alignments of these three disciplines with respect to the problem of addiction. In psychiatry and pharmacology, a mosaic resulted: each discipline carved out a problem set that was meaningful to it but did not engage questions of interest to the other. Since each discipline's cognitive construction of addiction gave the other free rein, a pair of coexisting models emerged, which appeared to complement each other. The important research on addiction in both of these disciplines was tightly linked to federal drug policy, which in the 1920s was crystallizing around supply-side control of drugs and the criminalization of addiction.

Sociologists were less successful in garnering resources for investigations of addiction in this period, but the roots of later ethnographic studies of addict subcultures are visible in the 1930s. Academic sociologists were rarely included in the funding and policy circles from which psychiatrists and pharmacologists studying addiction in the 1920s and 1930s were able to profit. Sociologists began to study addicts in noninstitutional settings, which psychiatrists and pharmacologists had not probed. Psychiatric and pharmacological formulations of addiction in this period were consistent with federal policy and were constructed in cooperation with the makers and implementers of that policy.

By contrast, early sociological studies of addiction formed the basis for a critique of that same policy. Study of social norms in drug-using groups formed part of the background for the development of symbolic interactionism, a theoretical approach that acknowledged the internal logic of subcultural norms and their meanings for individuals rather than portraying them simply as deviations from mainstream norms. This theoretical model informed a profusion of ethnographic studies of drug-using groups beginning in the late 1960s. From the 1970s on, policy makers came to value sociological data as policy input, and the federal government funded ethnographic studies through the National Institute on Drug Abuse. For example, ethnographic studies identified new trends in drug use; this information contributed to the planning of prevention and treatment methods, in addition to alerting enforcement bodies to changes in the marketing and use of drugs.

In the 1980s, the costs of marginalizing drug users from the mainstream medical and public health infrastructure became starkly clear when HIV infection rates soared in New York and other cities among both drug injectors and their sexual partners. Drug ethnographers benefited from the theoretical models and entrée into drug-using communities that they had already developed to probe the relationship of drug use to transmission of blood-borne infections. However, habits of thought and political structures long anchored in a drug war mentality blocked rapid government response to the implications of such research. In the vacuum created by government inaction, grassroots groups developed needle-exchange programs, not only to reduce transmission of HIV, but also to address the broader health needs of drug injectors. Innovations in HIV prevention across the board have shown the importance of including the voices and actions of those at risk of infection in planning and implementing public health responses to blood-borne infectious disease. Drug users have had to overcome decades of silencing and incarceration to claim their right to health care and public health protection.

As of this writing, epidemiologists have established the efficacy of needle-exchange programs in reducing HIV transmission, but rates of hepatitis C have not yielded even in localities with strong-needle exchange programs. The ability of the hepatitis C virus to survive in circumstances that effectively prevent HIV transmission has created the newest public health challenge for drug injectors. Developing effective prevention of hepatitis C requires precise knowledge of injection practices, which can only come from trusting partnerships between researchers and injectors.

Examining each of the disciplines that has studied addiction shows that historical specificity and contingency have been crucial in determining the individual settings in which the important research occurred and what its effects were. Despite the importance in psychiatry, pharmacology, and sociology of the same larger social themes, within each disciplinary setting, a complex interaction of disciplinary status, problem structures, individual actors, and settings led to unique research infrastructures and configurations of knowledge creation. At the same time, resolutions and accommodations within each of these disciplines created a larger interdisciplinary pattern, in which the cognitive and social space available to each was conditioned by concerns pertinent to the others. During the classic era of narcotic control, the result was a medical and scientific construction of opiate addicts as deviants with inherently flawed personalities in the grip of a physiological and psychological dependence from which few would ever escape. As the AIDS epidemic has revealed, both research and policy continue to be constrained by this view.

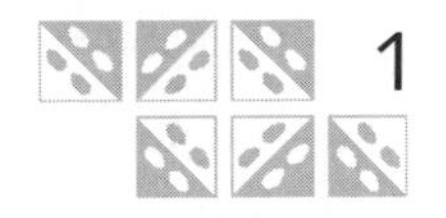

1

Heroin Addiction and Urban Vice Reform

In 1908, James Martin, then aged 21 and working in a Coney Island music hall, joined a fellow waiter on a double date with two women his co-worker had chatted up. Apparently wanting to impress his new friends, James suggested that they go at midnight to an opium den where he had entrée. None of the group had ever smoked opium before, but James had learned the complex skill of cooking up the substance for smoking while rooming with an opium smoker, and it was through this acquaintance that James was able to gain admission to the opium house. As James told it: "We smoked for two hours and I had intercourse with the girl and my testicles became swollen at that time and the pain was so severe that I kept smoking the stuff to relieve the pain. I stayed in that place for a week every night going back and forth and so I started my habit that way."[1]

After this first episode of opium use, with its mixture of social and medical motivations, James Martin smoked opium regularly for the next two years: "I would stay in the house for a week at a time while I had money." Initially, the expense was not much of a problem, because "a can of hop was only $4.00"; but "when the government made that law, it jumped up to about $50.00 a can." "That law" was the 1909 Smoking Opium Exclusion Act, which banned importation into the United States of opium for purposes of smoking.[2] Addicted to opium and facing dwindling supplies and rising prices, James switched to morphine, a more easily available drug, whose sale was still unregulated (except for the labeling requirements called for in the Pure Food and Drug Act of 1906). Then, he said, "somebody [in his social circle] came out with heroin so I was one of the first to use heroin." From sniffing heroin, he soon changed to injecting it subcutaneously to "get a better kick out of it."

This progression from milder to more intense forms of opiate use was a common pattern for a generation of young working-class men in crowded urban neighborhoods in cities like New York and Philadelphia who sampled opiates and became occasional, then regular users. Both their own rising tolerance to the drug's effects and the rising intolerance of policy makers resulted in reduced availability and rising prices. In response, users adopted more powerful forms of opiate (heroin rather than opium) and modes of delivery that intensified the power of a given dose (injecting rather than sniffing it). The psychiatrist Pearce Bailey, an astute observer of the emerging heroin scene in New York, described the pattern of switching to heroin as opium became less available and noted the former drug's advantages:

> In the old days, before the police shattered the opium joints, those who had become friendly with opium from social custom carried out their ceremonial by "laying in" for smoking the pipe. But with the disappearance of these resorts, old smokers and recruits also were obliged to turn to other forms of opium-taking, and [in about 1911] they deserted for the most part all other habit-forming drugs for a new derivative of opium. This new drug, which was heroin, won an immediate and widespread popularity. It had advantages over all rivals since the days of the pipe. It was cheap, it demanded neither layout [opium-smoking paraphernalia] nor hypodermic syringe, and could be taken for a long time without disturbing the health. It stopped the craving without diminishing working capacity to a degree which would prevent the earning of money to buy the drug, and last, but not least, as it is sniffed through the nose on a "quill," the addict could take it without much fear of being interfered with.[3]

James Martin had also sampled another drug in common use at parties and dance halls, cocaine: "At the time you could go in and get five cents worth in a drug store, very cheap at that time. Nobody knew what it was." Cocaine, though, did not suit James: "It made me weak, I was afraid to go out on the street." It made his heart beat so fast that he couldn't speak. Above all, James found that cocaine "makes you too uncontrolled. You can control yourself if you use the other stuff"—that is, opiates such as opium, morphine, and heroin. So he stopped using cocaine after trying it a few times at parties, but continued using opiates for at least eighteen years.

James Martin's drug habit affected his life in many ways. As his tolerance to the pleasurable effects of opiates rose, he needed to increase the doses he took; in time, he needed these higher doses just to sustain a sense of normality and stave off the sniffing, twitches, muscle cramps, vomiting, and diarrhea of withdrawal. In 1914, however, the passage of another law, the Harrison Narcotic Act, further restricted legal access to opiates. This law made possession or use of morphine or heroin illegal except as authorized by a physician.[4] Lacking a compelling medical condition that might warrant treatment with opiates, his sometimes painful testicles notwithstanding, James was now forced to obtain his drugs in the illicit market that sprang up in the wake of state laws and the new federal law restricting opiate sales. Under prohibition, the prices of morphine and heroin rose, just as the price of opium had done following its criminalization in 1909.

At times, James had turned down job opportunities that would have taken him away from his home in New York, where he had established his connections with drug sellers: "You could not go because you don't have enough stuff to last you and you would be afraid you could not connect and you could not go." Fear of arrest also deterred him from moving to a new location, where he would have to trust comparative strangers and might perhaps inadvertently buy drugs from an undercover narcotics agent. For six years, he always bought from the same man, also a user; this steady connection with a trusted seller spared him the risk of perhaps paying an unscrupulous dealer $40 "for a can of talcum powder." But, as he noted, there was also risk in staying put: "Neighbors see these fellows waiting around for the connection and they write a letter to . . . the Police Department." James had been arrested several times for drug possession; he once served a five-week jail sentence. "I guess they aint [*sic*] too many addicts that are not arrested," he observed. Caught between the need to satisfy his addiction and the strictures on his actions resulting from law, James organized much of his life around drug acquisition and drug use.

James Martin was a married man and the father of a 17-year-old son. His wife, Mona Flaherty, had been a young woman in his New York neighborhood who had taken a liking to him. James did not reciprocate her affection at the time, and he had "only had intercourse with her once or twice" before she told him she was pregnant. He initially refused to marry her, on the grounds that he did not have a trade and could not support her. Mona's aunt reported him to the authorities, however, and from prison, he agreed to marry Mona as a condition of his release. For the first two years, this was a marriage in name only, because James refused to live with Mona and continued his life of smoking opium and spending time with other women. Undeterred, Mona, as James put it, "stuck to me. . . . I could not seem to shake her somehow." Eventually, James began living with her, taking on steady work in order to support her. Over the years, his early regret over the marriage gave way to a deepening affection for his loyal wife.

Settling down did not stop James Martin's drug use. Rising prices and his own increasing doses saddled him with an ongoing and burdensome expense. As a young man, he was able to earn as much as $100 a week working as a stevedore, but age and, probably, the effects of syphilis, which he had contracted as a young man, had lessened his capacity for strenuous physical labor. By 1926, when he was 38, James earned $190 a month working as a bridge tender, raising and lowering bridges over which train cars passed from boats onto train tracks. His drug habit then cost him about $80 a month, or almost half of this salary. James supplemented his earnings playing poker, a game he was skillful enough at, by his own account, to win at most of the time; however, winning streaks were not consistent enough to keep him from falling into debt at times.

In his youth, James had hung out at the beach at Coney Island, gone to the fights, and taken in an occasional show, but over time, his world had contracted to the neighborhood where he could obtain his drugs. At 38, he rarely went out any more and no longer took his wife out in the evening as he had been accustomed to doing. His drug initiation had occurred in a social setting; as James put it: a "young fellow" would "get in with some company and . . . see them all enjoying themselves . . . at a party . . . and if you aint got a strong will power . . . then you take it once. Then if you get with them a week after that and take it again . . . you are hooked I guess." But after eighteen years of drug use, he had no close friends. Speaking of addicts, he said, "I don't think anybody else has in this line. . . . You don't mix with anybody else, you keep to yourself more or less." The drug had become increasingly central in his life: "When you get to be a drug fiend why the drug is more like a companion. You're contented as long as you have the drug. Well no," he amended, "it is not content. You are far from being happy." The pressures to earn enough to buy the drug formed one problem; the effort to evade arrest and actually obtain heroin, another. Furthermore, the drug no longer provided the pleasure it had earlier. Now, he said, "it's a curse all the way through. There's no pleasure to it. People think you get dreams and sweet thoughts and all such nonsense, but after you get addicted to the drug, there's nothing to it. You got to have it, if you don't you collapse."

James had sought relief several times by checking into treatment institutions, including Riverside Hospital in New York, but he had always reverted to drug use shortly after completing a course of treatment, which typically lasted seven to ten days. As he noted, in the 1910s, "there was a lot of experiment cures." By 1926, when James Martin entered the narcotics ward at Philadelphia General Hospital for another attempt at a cure, medical professionals were no longer experimenting widely with drug treatment. Twelve years after passage of the Harrison Act, addicts who wanted help in ending their drug habits had few recourses. The Jones-Miller Act of 1922 stiffened penalties for narcotics violations. The medical community responded in tandem. New York's Riverside Hospital had stopped accepting patients who had relapsed to drug use after previous treatment stays; Bellevue Hospital (where James Martin had been sent once for a cure following an arrest) and Metropolitan Hospital had both closed their drug wards.[5] Philadelphia's municipal hospital had one of the few remaining narcotics wards where patients like James Martin could check in for a round of the hyoscine treatment—a ten-day hospitalization in which the atropinelike deliriant hyoscine was administered during the withdrawal phase.

Addicts as Clinical Material

James Martin's account of his life as an addict has been preserved because, at Philadelphia General, he became a subject in a clinical research study intended to provide a better understanding of opiate addiction and how to treat it. The case notes from this research project provide a rare view of how the primarily urban, working-class addicts who began their opiate use in the first two decades of the twentieth century managed their habits and their lives, coped with stiffening legal sanctions and shrinking treatment resources, and created a repertoire of survival skills and a social world of clandestine opiate use that transmitted those skills to new users. This first American cohort of recreational opiate users also provided the "clinical material," in the common phrase of the day, for researchers eager to understand the physiological, psychological, and social nature of addiction. From the encounter between these addicts and the psychiatrists and other physicians, psychologists, and pharmacologists who examined them, at Philadelphia General and elsewhere, a view of addiction emerged that sustained a medical, scientific, and policy consensus for much of the twentieth century. For most professionals, these investigations validated the need for criminalization of opiate use and punishment of addicts.

In contrast, sociologists studying addicts in the 1930s developed the foundation for a challenge to that consensus, but this remained a minority voice in discussions of addiction well into the 1960s. From about the 1950s, other professional and academic groups added to a small but growing chorus of concern that addiction should be managed as a medical problem, not a criminal one. The spread of illicit drug use in the 1960s to include large numbers of middle-class youth increased the pressure to develop a less stigmatizing account of how people became addicted. In this context, liberal critics of drug prohibition argued that criminalizing opiate addiction had driven addicts to an underground market whose inflated drug prices forced them to commit crimes to earn enough to support their habits. These critics charged that the drug laws had created addicts whose criminality arose from the twin grips of intractable addiction and bad law. They offered an environmental explanation of addiction in reaction to the earlier, essentialist view that addicts were innately criminal, and that their flawed characters explained their addiction.[6]

David Courtwright's analysis of the demographics of opiate addiction between about 1880 and 1940 demonstrates that a shift from medical use of opiates, primarily by women, to recreational use, mostly by young working-class men, predated passage of the Harrison Narcotic Act.[7] This challenged liberal critics' argument that drug prohibition had in effect created the male criminal junkie who dominated public discourse. As Courtwright argues, laws like the Harrison Act were passed in part as a reaction against behavior that respectable observers saw as socially destructive and personally damaging.

The drug-using career of James Martin, and those of many of the addicts treated in the 1920s in Philadelphia General's narcotics ward, reveal the circumstances that gave rise to a new pattern of opiate addiction. Legislation did not single-handedly create the criminal addict; rather, the passage of laws prohibiting possession of opiates and cocaine was part of the larger context in which the first cohort of heroin addicts developed their social networks and survival skills. Other factors shaping the social milieu of the addict included the move to suppress prostitution and the alcohol prohibition movement (as well as less successful campaigns to ban other disapproved behaviors, including a movement to ban tobacco smoking).[8] In turn, such movements were responses to a burgeoning new entertainment scene in immigrant, working-class neighborhoods in America's industrializing cities, in which the traditional boundaries between respectability and vice were blurred. Together, urban reformers and thousands of addicts like James Martin created the perception of the junkie prevalent in the United States in the twentieth century and still today. In smaller numbers, and with much less visibility in official sources, women, too, figured in this emerging scene. The junkie emerged from the turbulent context of the transformation of vice in the American city between

1890 and 1930. Narcotics reform simply made drug users' lives harder and more dangerous.[9]

As James Martin's occasional need to adjust his drug use to the exigencies of the newly passed laws suggests, his addiction occurred at a time of rising national concern about the use of opiates. As America industrialized and urbanized, immigrants flowed into its cities from southern and eastern Europe to take jobs in factories and in the retail and amusement venues that were becoming increasingly prominent features of American downtowns. Social life in working-class urban neighborhoods was profoundly transformed, provoking a variety of responses from older middle-class and elite groups, including the settlement house movement and an anti-vice reform movement, which centered its concerns on prostitution but also agitated to prohibit the sale of narcotics. Both of these currents of social reform contributed to framing opiate addiction as a problem to be studied by laboratory scientists, social scientists, and physicians.

The first cohort of recreational opiate users emerged at the turn of the twentieth century, when tolerated, segregated vice districts flourished in many American cities. The existence of these vice districts was supported, not only by a brisk demand for their goods and services, but also by a complex set of relationships with law enforcement, local political structures, and more respectable sectors of the local economy, including the owners of buildings where brothels operated and hotels that housed traveling men, who learned from bellboys where to find local brothels.[10] As the anti-vice crusades gained steam, they achieved victories that transformed the ecology of urban vice, although not always in ways anticipated by the reformers. These included the closing of red-light districts, suppression of prostitution, the prohibition, for a while at least, of alcohol, and an enduring ban on the sale of opiates and cocaine except as carefully regulated medicines. Those who had begun opiate use before about 1910 adjusted their use as the market for opiates underwent profound alteration, and those who began using after implementation of the Harrison Act had to learn quickly how to navigate an illicit market. The case of Daniel Hood illustrates the relationship of opiate use to other currents shaping the world of urban amusements and vice in the Progressive Era.[11]

Coming of Age in the City

Daniel Hood was an alias adopted by Daniel Schultz, born in Philadelphia in 1898 to parents who were also native Philadelphians. The family lived in a working-class neighborhood within a dozen blocks of City Hall and of the city's Tenderloin district. Daniel's father, an electrician and jack-of-all-trades, was an alcoholic who beat his children. By age 12, in 1910, Daniel had begun running away from home, at least for short periods. At 14, in 1912, he quit school and began running with a gang.[12] He spent time drinking with fellows who hung around the Franklin Race Track. Dropping out of school, hanging out with friends at the racetrack, drinking, and running away became for Daniel a pattern that helped determine the course his adult life would take.

The same year that Daniel dropped out of school, 1912, Philadelphia's reform mayor Rudolph Blankenburg appointed a vice commission to investigate the problem of prostitution in Philadelphia. The commission's investigation was carried out under the supervision of George Kneeland, who had recently completed an investigation of prostitution in Chicago and was simultaneously directing such an investigation in New York.[13] The Philadelphia vice commission's report appeared in 1913, and the similarity of its findings to those of the Chicago and New York reports reflected not just Kneeland's involvement in all three but the concerns of vice reformers nationally. Across the United States, social reformers in major cities were getting vice commissions appointed to examine the problem of prostitution; these produced a similar melange of conclusions and recommendations. Although these commissions acknowledged that prostitution arose from a complex set of causes, including the low wages paid to women in factories and retail stores and the political and policing arrangements that supported what Lawrence Friedman calls "the Victorian compromise," which allowed prostitution to flourish in tolerated red light districts, despite laws against it, as long as it was kept spatially and socially separate from the respectable middle-class city,[14] they also emphasized the overdeveloped sexuality and blunted moral sense they attributed to most prostitutes. "They are bad because mentally they are unfit to be good," the Philadelphia vice commission declared.[15] Unanimously and without reservation, they recommended complete legal suppression of prostitution; in so doing, they rejected any form of regulated prostitution as practiced in several European countries.[16] As states and municipalities passed new, more severe laws against prostitution, and as cities elected reform mayors to replace machine politicians who had tacitly allowed prostitution, local police were instructed to close down brothels once and for all.

Unlike some other cities, Philadelphia retained its vice commission after the mayor had begun to close down brothels and other sites of prostitution, which were centered in the Tenderloin district, which included the city's Chinatown.[17] As the vice commission noted, Philadelphia had never had a strictly segregated red light district; rather, brothels and other amusement sites typical of Tenderloin neighborhoods existed in an area also populated by "respectable people." Addicts like Daniel Hood arose from this mix of vice, commercial amusements, and working-class neighborhoods; as their drug involvement deepened, their social worlds included the conventional worlds of work and family and the hustling world of the street scene.

"Conditions were in a turmoil; prostitutes and madams were leaving town or moving to other parts of town, and all the pursuers of the vice business were agitated and in a state of great unrest," as a result of Mayor Blankenburg's crackdown, the vice commission noted.[18] This turmoil was felt at all levels in the networks supporting prostitution, from the madams who managed brothels down to the teenage messenger boys who received an early education in the sporting life as they guided customers to locales where they would meet prostitutes, delivered chop suey to prostitutes' rooms, and occasionally arranged bail for jailed prostitutes. Philadelphia vice commission investigators' interviews with several such messenger boys show the impact of vice reform on the world of urban amusements that Daniel Hood and his friends roamed.

Messenger boys who worked for companies that provided service to brothels were comparable to hotel bellboys, cabdrivers, and others whose occupations called for knowledge of the landscape of urban vice. Such workers depended on their own hustle to maximize their earnings through fares, fees for specific jobs, or tips. In this way, they differed little from salesmen, small-scale entrepreneurs, and countless others who sought to improve their standard of living by deploying energy and savvy in a fast-changing urban world rather than through repetitive labor on a fast-moving assembly line. Certain skills, though, were especially important for those linked to the marketing of vice just as those markets were undergoing the upheaval engendered by Progressive initiatives to make commercial sex and psychoactive drugs illegal. As commercial sex was forced underground, and operations were moved to evade police discovery, guides to urban vice had to remember passwords, maintain personal connections, and keep up with shifting locales. Changes in vice enforcement upped the premium on being able to read people quickly, almost instinctively, to distinguish customers from undercover officers.

The main impact of the suppression of prostitution was to make life harder on all involved (except, presumably, the reformers and the enforcers). One messenger boy reported that in the immediate aftermath of Blankenburg's call to close down the brothels, he "had lots of calls to take suit-cases from the houses to the railroad stations. The women were going out of town."[19] Those prostitutes who stayed in Philadelphia were pushed out of the brothels where they had shared their lives with other women under the supervision of a madam who perhaps combined financial exploitation and sympathetic treatment; they were increasingly likely to be managed by pimps. They lost the social supports provided by living in close relationships with other prostitutes and were more vulnerable to exploitation and violence.[20]

Faced with decreasing revenues from prostitution, all parties sought alternative ways of increasing their incomes. Prostitutes colluded with their pimps in what messenger boys described as "panel houses." Here, the prostitute laid her customer's trousers near a wall panel. While she engaged the customer in sex, her partner opened the panel and slipped the customer's wallet out of its pocket. As one messenger boy put it, "It's not safe to go into any whorehouse in the city now."[21] Once brothels lost their geographic stability and their relatively stable clientele, what little protection patrons and prostitutes had enjoyed against each other disappeared.

Pimps also turned to selling opium and cocaine, a business that would become increasingly lucrative as laws prohibiting these drugs were enacted and enforced. In 1913, when the vice commission's report was published, opium intended for smoking had been banned by federal law since 1909. Although the Harrison Act had not yet been passed, numerous states and localities had passed laws against possession of opiates and cocaine; among the recommendations of the Philadelphia vice commission was a call for state legislation to make cocaine possession a felony rather than a misdemeanor, as current law provided for.[22]

Messenger boys' accounts illustrate the ways that vice markets existed as parts of working-class neighborhoods, providing jobs for young adolescents who also faced a range of choices about livelihood and social connectedness. One boy lamented that the current turmoil had cut into his ability to contribute support to his parental home. "Last year," he reported, "I used to make three and four dollars a night in tips alone. I used to give my whole wages to my mother and keep the tips." His immediate segue into a description of the "French houses" where "they would take a man up any way he likes" suggests the police crackdown had also compromised his ability to balance the life of a young man still living with his parents with learning the ways of the street.[23] Another boy turned down an opportunity to become a pimp. "I had a chance to put a young girl to work for me," he said. "She asked me if I wanted to solicit business for her on the street while she was in a room. She did not want to go on the street, because she was afraid of meeting people she knew." But this boy was not ready to break with his family: "I knew it meant living with her all the time. My parents would kill me if I did not come home every day."[24] He also expressed fear that other pimps would retaliate against his incursion into their turf. Similarly, two boys expressed reluctance to replace lost income by selling drugs, as they reported many pimps had done, because of tight police surveillance under the Blankenburg regime.[25]

How these boys resolved the conflicting pulls of family ties and street temptation is not recorded, but Daniel Hood's career illustrates how drug use in a "reformed" urban context contributed to his falling away from a respectable life and entering more deeply into the underground world that supported drug use. In about 1913, when he was 15, like countless other teenage boys on the streets of America's large cities, Daniel began sniffing heroin from time to time. His first use occurred when a friend offered him some heroin, saying "this would fix me up after drinking." That his young friend offered him heroin rather than opium is at least in part explained by the passage in 1909 of the act that forbade importation to the United States of opium for smoking. Daniel thus began using heroin just when that more powerful opiate was replacing recreational opium smoking. At one point, Daniel went three months without using heroin; but after his father died (when Daniel was still 15), he "took it again blowing it, and after I blowed it I felt better." Occasional use drifted into regular use, until Daniel came to prefer heroin to whiskey.

At some point, Daniel "went to the country for a week." There, he began to feel miserable, but by the sixth day, he felt better. He returned to Philadelphia and described the experience to one of his heroin-using friends, who interpreted it for him: "He told me I had a habit and he fixed me up." For another several months, Daniel continued regular use, but in his view, he was using heroin as a medicine to deal with the aftereffects of drinking. In saying this to the psychiatrist who was interviewing him, Daniel was probably not simply rationalizing the origins of his addiction to a middle-class professional with control over his treatment. Morphine had been commonly prescribed to ease hangovers and alcohol withdrawal for some decades, and his view was consistent with the blurred boundary between medical use of a psychoactive drug and what was increasingly being characterized as "vicious" use (i.e., use that constituted a vice).

When a trip to Burlington, New Jersey, again interrupted Daniel's regular heroin use, he once more went through several days of misery.

Apparently, he was trying to stop using the drug, because he said he had thrown away his supply when he left for New Jersey. However, although his withdrawal symptoms again abated around the sixth day, his discomfort did not end. Nothing felt good; even a "cigarette tasted like poison." He sought relief in a dose of heroin, and "it took just a minute before I was fine again."

Daniel became a heroin user not only in the midst of the Philadelphia cleanup campaign but just at the time when reformers were pressing for passage of the Harrison Narcotic Bill. Brought before Congress in 1912, the bill failed, but it passed in 1914. The Harrison Act was a milestone in several respects. As the first federal law to prohibit possession of a drug, it was a precursor to the national experiment with alcohol prohibition that would begin a few years later. As an example of Progressive Era regulatory legislation, it was part of the expansion of the federal role in regulation. Some of its strongest opponents on the floor of Congress argued that, by regulating an aspect of medical practice, the bill threatened to usurp a recognized area of states' rights, the right to regulate the practice of medicine. Nevertheless, the American Medical Association, despite its role in protecting the prerogatives of the practicing physician, supported the bill as part of its efforts to dissociate physicians from the charge of causing iatrogenic addiction. Like many of his contemporaries who sniffed heroin, Daniel found that implementation of the Harrison Act after March 1, 1915, made the drug scarcer and more expensive. Meanwhile, no doubt, his continued regular use was raising his tolerance for his accustomed dose. These factors jointly led him to change how he procured the drug ("I got the stuff off a peddler, on the street, when the law went into effect") and how he administered it ("I blowed it until the law went into effect and changed to the hypodermic").

At 14, in 1912, Daniel had run away to New York and learned bookbinding at a trade school. As his drug use escalated, Daniel found he could not hold down a job as a bookbinder, and he shifted to other methods of supporting himself. These ranged from the quasi-legal to the patently criminal. He scalped tickets and developed a pattern of theater work in the winter and traveling with a carnival in the summer, running a spinning-wheel game. Moving back to Philadelphia, he worked for one or more spells as a cabdriver and as a messenger for an employment service—both jobs intimately linked with the marketing of vice, as we have seen. At one point, he joined his brother in a scheme to sell fake diamond rings. He served some time in jail for possession of drugs. He learned the subterfuges that addicts developed to hide their addiction; thus, at his first arrest, faced with a spell in jail, where he would not be able to obtain drugs, he hid some heroin under a sticking plaster, which he claimed covered an abscess (a common complication of repeated injection of drugs).

Drug use could make work difficult in the illicit labor sector as well. For a year, Daniel ran a crap game in Philadelphia, apparently during a period of abstinence; but, he said, he "then went on this stuff and got careless, so I stopped. You had to run them square," he said of the games where professional gamblers played, "or you would get killed." Both as a buyer of heroin on the illicit market and as an individual supporting himself through carnival work and confidence games, Daniel needed to be able to read people instantly, to adapt quickly to changes of location, personnel, and passwords, and to manage carefully that which could be disclosed and that which must remain hidden in any particular setting. Addicts like Daniel thrived in the quasi-legal corners of American cities, in a culture that admired "hustle" as a way of getting ahead, even as it disdained the source of their ambition and condemned the means by which they supported themselves and their habits.

Sex, Drugs, and Vice Reform

For the vice commissions, an investigation of prostitution led to identification of drug use (opium, heroin, and cocaine) as a related social problem. For Progressive reformers, prostitution and drug use formed part of a constellation of urban vice, and their understanding of this set of problems helped create the context for scholarly study of them. The vice commission reports reveal three themes linked to the Progressives' interest in environmental and systemic explanations, the larger urban and industrial transformation America was undergoing, and the rising authority of science and medicine to explain problems related to sex and drug use. Despite the academic or technical tone of much of the discussion, it remained tinged with a lingering set of moral beliefs about the wrongness of vice and the individual culpability of those who engaged in it.

First, Progressive reformers imposed a rigid dichotomous moral standard on a flux of related and disparate behaviors; in doing so, they blinded themselves to the fluidness of the boundaries separating legitimate from illegitimate behaviors in the eyes of many denizens of working-class neighborhoods. The Philadelphia vice commission, for example, condemned not only the outright exchange of sex for money but also sexual practices they considered perverted, any form of homosexual contact, dancing that involved close physical contact and sensuous movements, and risqué theater performances, including burlesque shows. They rightly observed that these were all part of a heterogeneous context of urban amusement and vice, and they strove suppress them all (while urging sanitized or chaperoned versions of such activities as dancing and stage performance) in the name of a middle-class standard of decency. By securing legal prohibition of a range of activities, they marked out an absolute boundary between what was to be considered legitimate and what was to be condemned.

Succeeding generations of Americans, however, would liberalize their sexual norms, their acceptance of unchaperoned socializing between young, unmarried men and women, and their tolerance for risqué displays on stage and screen. Repeal followed Prohibition of alcohol, bringing with it social tolerance of extravagant levels of drinking. In many ways, the boundary defining deviance shifted in a liberal direction. One result, however, would be a deeper chasm separating those still considered deviant, even by these more liberal standards, from conventional acceptance. Prostitutes, drug addicts and homosexuals would, in the ensuing decades, be left on the wrong side of that chasm, and paths back to conventional social roles would be fewer and more tenuous.

Second, the vice commissions helped create the context for the professionalization of concern about vice that had formerly been framed as charitable activity. Such professionalization included the rise of social work as a profession and the framing of social problems as objects of scholarly study in the emerging or transforming disciplines of the social sciences. In the broadest sense, Progressives' attempts to understand the phenomenon of prostitution mirrored social scientists' concerns about the transformation of an industrializing and urbanizing America. The latter attempted to reconcile what they saw as America's unique political heritage with the rise of labor activism and the potential for class tension that, in European settings, gave rise to revolutions and socialist movements.[26]

More specifically, the vice commissions' attention to structural and spatial issues anticipated the approaches of sociologists, especially those of the Chicago School. The commission reports discussed the economic prospects young working-class women faced, contrasting working in factories or stores with prostitution. They traced the connections between prostitution and legitimate sectors of economic activity, and blamed landlords and businessmen who profited, if indirectly, from the sale of sex. They pinpointed the locales where prostitution occurred, including dance halls, bars with rooms where prostitutes worked, and cafes where prostitutes could pick up customers. Although the vice commissions ultimately rejected structural explanations, blamed women who became prostitutes, and urged the suppression of prostitution, their wide-ranging inquiries produced many data of use to later scholars. One of these scholars, the Chicago-trained sociologist Walter Reckless, used the Chicago vice commission's 1911 report as the basis for assessing the changes in patterns of prostitution since the move to suppress it had driven it underground.[27]

Third, Progressive reformers identified venereal disease, syphilis, in particular, as a social problem arising from prostitution that threatened the middle-class family. Referring to syphilis's long clinical course and involvement in many organ systems, Sir William Osler, the eminent Canadian clinician, famously said, "Know syphilis in all its manifestations and relations, and all other things clinical will be added unto you."[28] One might add that following the course of syphilis through the various reform currents of the Progressive Era can similarly illuminate a historical movement. New scientific and clinical understandings of syphilis coincided with rising concern about the social costs of venereal disease. Social hygienists placed syphilis (and to a lesser degree, gonorrhea) at the center of a complex of problems they saw as eroding the health and fitness of the American family. In this construction, prostitutes harbored the disease and transmitted it via commercial sex encounters to husbands (or future husbands), who then brought the disease into the sanctity of the home. Infected wives, in turn, gave birth to infected, sickly infants, or they became barren as an effect of the disease. This threat struck at the heart of elite groups' concerns that declining middle-class fertility was vitiating America's Anglo-Saxon heritage and national vitality.[29]

Alcohol and drugs were also prominent features of the social scene that seemed to give rise to prostitution and syphilis.[30] Together, these issues straddled the boundary between social and scientific arenas. As problems, drug addiction and syphilis shared many attributes. Both were seen as originating in the same seedy social world of dance halls, bars, and poolrooms. Both were understood to be contagious, transmitted from one person to another by vice. Both were framed as diseases to be studied by science. Both were linked to the development of new medications. Some physicians and reformers saw the rising prevalence of addiction as a by-product of the scientific quest for improved medicines: morphine was the first alkaloid to be isolated from its plant source, and heroin, first marketed in 1898, was the result of pharmaceutical research. Syphilis was the target of Salvarsan, the "magic bullet" that Paul Ehrlich launched as an early triumph of drug-development research. Both addiction and venereal disease were increasingly framed, from the 1890s on, as public health problems, and both would become important objects of concern for the U.S. Public Health Service in the twentieth century.

Syphilis differed crucially from drug addiction, however; while physicians, scientists, reformers, and politicians debated whether addiction was a true disease or simply a moral failing, the explication of syphilis as a bacterial infection in 1905 brought it into the bacteriological revolution, with all its ontological authority. August von Wassermann's development of a complement-fixation test for syphilis the following year gave clinicians an unambiguous diagnostic tool for identifying sufferers from syphilis, while addiction continued to bedevil diagnosticians.

An important early impact of the ability to determine at autopsy whether an individual had contracted syphilis during life was the discovery that many of the patients who had lived out their last years as inmates of insane asylums evidenced syphilitic lesions of the central nervous system. The syndrome known as general paresis was now understood to be a late-stage, chronic manifestation of syphilis, and living patients could be given Wassermann tests to confirm the diagnosis.[31] Some observers estimated that 20 percent of asylum patients owed their mental illness to syphilis.[32] For a while, this finding sparked hope among psychiatrists that preventing syphilis would substantially reduce the incidence of mental illness, and that perhaps other forms of insanity would be found to have bacterial causes. This optimism faded—the new understanding of syphilis did not auger a bacteriological revolution in the study of mental disease—and its identity as a bacterial disease hardly eliminated moralistic overtones from scientific and policy discussions regarding venereal disease control.[33] Nevertheless, over the course of the twentieth century, especially following the widespread use of penicillin to treat syphilis, it was brought substantially under control, and cases rates in many cities were extremely low by the 1970s. In 1981, when AIDS appeared on the scene and was quickly understood to be both sexually transmitted and connected to heroin injection,[34] drug use and sexually transmitted disease again formed a tangle of related problems with urgent policy dimensions.

Young men like Daniel Hood learned their way around the city just as the Progressive Era forces described above were creating continuing flux. This period witnessed a cohort of first-generation American children creating an urban street culture in which their own immigrant parents' traditions were no useful guide and they had to find their own way.[35] The lack of adult guidance, the social complexities of the city, the changes wrought by vice reformers, all placed a premium on quick learning on the street. As one observer noted, "The street educates with fatal precision."[36] Street smarts became essential to navigate the markets for vice as they were increasingly driven underground. In this setting, junkie culture was forged. It emerged not as a pharmacologically determined set of behaviors but as an adaptive response to a rapidly shifting social context of use—a social context whose contours were most fundamentally shaped by the nature of the drug marketplace, which, as a result of changes in law and medical practice, was increasingly confined to the illicit sectors where vice was purveyed and crime flourished.

The pattern of addiction for women was somewhat different and is harder to trace. Certainly, the number of female opiate addicts was decreasing over the period from about 1900 to 1930; fewer women became medically addicted as physicians reduced their prescription of morphine and as restrictions on pharmacy sales, culminating in the implementation of the Harrison Act, reduced availability. With access to opiates restricted, women struggling to deal with their addiction faced a narrowing set of options.[37] For some, sympathetic physicians continued to prescribe morphine despite the heightened legal risks of doing so. Others undoubtedly gave up using morphine. Some whose cases appear in the records of the Philadelphia General narcotics ward maintained their addiction by learning to acquire drugs under the new regime of prohibition. A tantalizingly brief record describes Gloria Fanelli's adaptation.[38] In 1898, when she began menstruating at age 12, her physician had prescribed morphine for menstrual pain. Taking morphine for menstrual discomfort was common in the nineteenth century, and the practice was sometimes passed down from mother to daughter. Some women became addicted as they progressed from monthly to more frequent use. Gloria developed a hypochondriac strategy to maintain her addiction. Learning to vomit at will, she persuaded physicians that she needed morphine to calm her stomach. Undoubtedly, more than just a simple desire to obtain drugs underlay Gloria's behavior, but it is likely that many addicted women sought to manufacture ongoing medical complaints to maintain a supply of drugs, and some must have found it necessary to move from physician to physician. For doctors, then as now, prescribing psychoactive drugs was a way of dealing with patients with vague complaints.

Women also began using opiates recreationally. Although males predominated in this category, numerous accounts in the Philadelphia General records discuss settings where men and women used opiates together socially, and some men described being initiated into opiate use by women. Between 1925 and 1928, inclusive, of 1,143 patients admitted to the narcotics wards (male and female) at Philadelphia General, 796 were men and 347 were women.[39] It is impossible to know to what extent this ratio was representative and to what extent women were more reluctant to seek treatment, or found treatment harder to get, because fewer facilities existed for them (Philadelphia General, for example, had three men's narcotics wards but just one for women).

Like Gloria Fanelli, Mary Garfield was also introduced to morphine by a physician, but under very different circumstances.[40] In 1914, when she was 17, a physician spotted her on the street and persuaded her to take a drive with him. Other drives followed, and soon the couple began going to parties together. After a time, the physician separated from his wife and moved Mary into his home; she began helping with his practice, including keeping his books. Soon thereafter, she discovered that he was a regular user of morphine. She made some attempt to dissuade him from using it, but then allowed him to give her injections of the drug. At some point, she traveled from their home in Lynn, Massachusetts, to nearby Malden for a weekend. On the Sunday morning, feeling "sick and miserable," she telephoned the physician, who told her to take a taxi home at once. On her arrival, he explained to her that she had a habit and relieved her discomfort with an injection; she was "soon her normal self again."

Mary was very much in love with this physician. Although he did not pay her for her bookkeeping work and was still legally married, he gave her a home and bought her clothes and whatever else she wanted. The two went on vacations together, and he often introduced her as his wife. They socialized with other drug users, attending parties where cocaine was used. After eight years of this life, the physician suddenly began exhibiting strange behavior. He became forgetful and despondent; he talked of suicide. He was admitted to the Danvers State Hospital at Danvers, Massachusetts, and died there three months later.

Left on her own, Mary returned to her former work as a stitcher in a shoe factory; in time, she became a shop "forelady." Forced now to acquire her own drugs, she switched between morphine and heroin, depending on what she could find. She stopped using cocaine because she could no longer afford to keep up with the social set who gave the cocaine parties she and the physician had attended. Some six months after the physician's death, she became "disgusted with the habit"; in addition, "the drugs were getting hard to get and 'it was difficult to trust people to get it for you.'" She voluntarily entered the Bridgewater prison in Brockton, Massachusetts for fifteen days, during which time she "broke cold turkey."

A man Mary had met at a dance hall began courting her, and although she did not love him, she married him. For two and a half years, she used no opiates, but she and her husband drank "occasionally." Mary missed her dead lover every day, and tensions emerged in her marriage to a man she professed not to be in love with, although she said he was in love with her. Nevertheless, Mary reported, "She feared he was rather gay, so they broke up their partnership and she, depressed, went back to the drug." Mary did not specify whether she was referring to his sexual orientation. The centrality of this issue to their breakup suggests so, but the point was not pursued. The interviewing psychiatrist consistently discussed homosexuality only obliquely, either describing certain male patients as effeminate or noting perfunctorily the absence of "perverted relations" in male patients' sexual histories.

They argued and separated; depressed, she resumed morphine use. Her husband sought a reconciliation but was discouraged to learn she was using drugs. Mary returned to her marriage, continuing to use drugs, but her husband's concern prompted her to seek treatment again, this time at Philadelphia General Hospital.

This, in any case, was Mary's account of her life and her addiction. As a set of recorded responses to a psychiatrist's interview questions, it may conceal as much as it reveals. At 17, by this account, she worked in a factory as a shoe stitcher and was willing to take an automobile ride with a presumably charming and apparently affluent stranger. Did Mary occasionally supplement her factory wages through prostitution? Was she drifting from factory work to the higher earnings she could get as a full-time prostitute, or had she already completed such a trajectory? Or was she simply a young woman easily persuaded to give up factory work for the excitement and greater security of living with a man who treated her well, exposed her to a smart party set, and gave her work (accounting) that may have been more satisfying than factory work? When she met her second husband at a dance hall, was she a taxi dancer there, dancing with men in exchange for five- or ten-cent tickets? Or was she simply out having a good time?[41] We don't know; despite her willingness to talk about her status as a mistress (the relationship, after all, became deeply important to her), she might well have held back any involvement she had in prostitution or dance hall work. We do know that, somewhere along the continuum from young women who socialized freely with men in settings that involved drug use and dancing to women who worked full-time as prostitutes, Mary became addicted to morphine, and this addiction posed problems for her.

James Martin, Daniel Hood, and Mary Garfield were in some ways typical of men and women addicted to opiates in the 1910s and 1920s, although in other ways each was unique. All began drug use as young adults in a social setting with friends who used drugs. Each moved from an early period of pleasurable involvement with the drug to the realization that he or she had developed a habit—and this knowledge was consolidated when more experienced addicts interpreted their first withdrawal syndromes as evidence of a habit. All adjusted their drug use at times in response to changes in the market—changes, in turn, affected by passage of drug laws. All came to wish themselves rid of their habits, and all had sought treatment more than once. All continued their drug use well after implementation of the Harrison Act on March 1, 1915. In numerous ways, they typified those who began using opiates before or just as the nation adopted a policy of prohibition of these drugs, and who developed more or less successful strategies for coping with the vagaries of the black market, the surveillance and incarceration mandated by law, and the efforts of professionals to help them get better. And their bodies and their life stories provided some of the raw material for researchers' constructions of opiate addiction. By the late 1920s, when James Martin, Daniel Hood, and Mary Garfield were studied at Philadelphia General, ideas about opiate addiction were coalescing around a harshly pessimistic view that fundamental defects of character underlay intractable addiction; but at the time they began their drug-using careers, in 1908, 1913, and 1914 respectively, a broader range of views still existed.

Theory and Practice Before 1920

The addiction model that became so influential by 1940 was not based on brand-new ideas; rather, a mosaic of disciplinary, policy, and institutional concerns created a context in which a particular set of already current ideas displaced others and gained dominance. The heterodoxy that characterized the pre-1920 period provides the immediate background for the consolidating work of researchers in the 1920s and 1930s. The next sections of this chapter survey ideas about opiate addiction in the following arenas: the legislative and policy activities of the federal government; the reform activities of the American Medical Association; and the research of various scientific investigators interested in opiates.

The Policy Context

Progressive Era reformers included opiate addiction among the broad range of social concerns they proposed to address with legislative and regulatory remedies. In the unregulated drug market of the nineteenth century, opiates were freely available in a wide assortment of medications marketed to the public. Opiates' power to soothe pain and to alleviate cough and diarrhea made them important therapeutic tools, and doctors prescribed them for a wide range of medical indications.[42]

Opiate addiction posed a direct threat to the Progressive ideal of the rational, public-spirited American citizen. Samuel Hopkins Adams, in a 1905 series of articles detailing the frauds perpetrated by unscrupulous drug merchants, called opium "the most dangerous of all quack medicines." By creating "enslaving appetites," opium and its derivatives dulled rational faculties and sapped self-control.[43] In part as a result of Adams's articles, which appeared in *Collier's* magazine and were reprinted and widely distributed by the American Medical Association, Congress in 1906 passed the Pure Food and Drug Act. Its provisions included the requirement that the presence and quantity of opiates in any preparation be stated on the label. Enforcement of the law proved more difficult than had been anticipated, and manufacturers won court victories based on skillful legal defenses.[44]

In 1914, the passage of the Harrison Narcotic Act marked the beginning of federal attempts to exercise direct control over access to drugs in the United States. Since about 1900, states and localities had enacted laws regulating or forbidding sale for nonmedical use of opiates, cocaine, and a few other drugs, such as chloral hydrate.[45] These laws reflected the traditional arrangement that the states, not the federal government, had responsibility for regulating the practice of medicine. Thus, the Harrison Act exemplified the expanding regulatory powers taken on by the federal government in the Progressive Era. Ostensibly passed as a taxation measure, with paperwork provisions to track distribution of opiates and some other drugs, the law allowed only physicians and pharmacists to dispense opium, morphine, or heroin to the public.

The Harrison Act was proposed largely in response to lobbying efforts by Secretary of State William Jennings Bryan and Charles H. Brent, head of a War Department commission of inquiry into narcotics laws under the Spanish in the Philippine Islands. These men, along with missionaries and traders concerned about the British opium trade in China, urged that the United States join a growing world movement to regulate the international trade in opium. U.S. delegates to international conferences on the world opium problem (Shanghai, 1909; The Hague, 1911) were in a poor position to advocate reform elsewhere when their own country lacked national legislation regulating the domestic opium trade. Congressional hearings on the Harrison bill concentrated largely on these international issues rather than on the evils of opiate addiction.[46]

Besides the provisions for regulating interstate shipment of opiates, the law also contained language prohibiting possession of certain drugs, including opiates (chiefly morphine and heroin) and cocaine by anyone without a license. The Harrison Act ended a period in which any substance could be packaged and sold as a medicine to the general public. Licenses to possess opiates were to be granted primarily to pharmacists and physicians, and members of the public could thereafter obtain opiates only through a doctor's prescription. Two forms of access to opiates were thus cut off: the drugs were no longer freely available in medications anyone could purchase; and legal supplies for recreational use were no longer available. (Passage of the Smoking Opium Exclusion Act of 1909 had targeted a form of opiate perceived as having no medical use.)

Because of the law's tax provisions and its focus on interstate commerce, the Treasury Department was charged with its enforcement. In the years immediately following passage of the law, federal and local authorities, as well as physicians treating patients addicted to opium, morphine, or heroin, grappled with the propriety of allowing addicts to continue to receive the drug they were addicted to without aiming to end their drug use or "cure" them of addiction. This procedure is called "maintenance" because the addiction is "maintained" through the continued administration of the drug. Although the law did not address this issue directly, the Treasury Department interpreted the legislation as prohibiting any form of maintenance, whether by public clinics or private physicians. The Treasury Department made allowance for physicians to maintain the addiction of chronically ill or elderly patients on the grounds that they needed morphine to treat pain or that undergoing withdrawal would be dangerous for them. Beyond such cases, Treasury Department officials equated maintenance with drug peddling, contradicting the view of some physicians that maintenance was an acceptable form of treatment that allowed patients to avoid the upheavals of withdrawal and the likelihood of a relapse.[47]

On the local front, however, many municipal health officials became concerned that the sudden lack of availability of opiates to the general public would cause addicts to go simultaneously into sudden withdrawal. There were fears both that people would suffer needlessly (after all, they had become addicted before the existence of a law proscribing their drug use) and that some would become violent or resort to crime in the attempt to secure new supplies of drugs. For these reasons, several cities and states established clinics to offer treatment services to addicts. There was confusion as to the mission of these clinics: were they to perform just a detoxification function, offering addicts steadily decreasing doses of opiates to reduce the severity of withdrawal, or could they also maintain addicts by supplying steady doses of opiates on a long-term basis, without the aim of achieving abstinence?[48] Regardless, as described below, the Treasury Department took a dim view of their activity.

One effect of the clinics was to group addicts in a manner visible to officialdom for the first time. Before the passage of the Harrison Act, individuals could obtain opiates from physicians, in patent medications, or through nonmedical channels (for example, buying opium for smoking or chewing until 1909). Possession of opiates or addiction to them did not in itself bring one to the attention of any authorities, and few public institutions existed to deal only with addicts. Private clinics offered treatment for addiction, and private physicians might either treat a patient for the symptoms of withdrawal or supply opiates on a steady basis; in neither case did the patient come to the attention of any enforcement body. Furthermore, addiction occurred among many segments of the general population and was not concentrated in any single group.

As addicts were registered at clinics throughout the country, they were questioned as to their past use of opiates (in part to establish a level of addiction so as to determine the appropriate starting dose at the clinic). Statistical profiles of addicts, including age and sex distributions, and age at first drug use and onset of addiction, became possible for the first time. Attempts to determine the prevalence of addiction in the 1920s were based in part on figures collected by the clinics. For the first time, such studies appeared to have a scientific foundation.[49]

Many private physicians believed that, under the provisions of the Harrison Act, they were free to prescribe opiates entirely as they saw fit. Some undertook to treat addicts by prescribing opiates for them on a maintenance basis. These included conscientious practitioners acting from a variety of motives: the patients were old or chronically ill, or the physician wanted to enable an addict to prevent withdrawal and avoid public exposure. Others prescribed opiates freely to all comers, sometimes without even the pretense of a physical examination.

The Treasury Department quickly began prosecuting physicians who were believed to be prescribing opiates improperly. Several appeals reached the Supreme Court, which was faced with interpreting the vague language of the Harrison Act. While earlier case opinions appeared to leave physicians considerable latitude, a pair of decisions in 1919 upheld the convictions of doctors who had prescribed opiates to addicts, stating that such behavior fell outside the bounds of proper professional practice. The Treasury Department officials in charge of enforcing the Harrison Act saw in this language a mandate to prohibit any form of addiction maintenance.

In the early 1920s, the Treasury Department moved against the public clinics that had opened to treat addicts, charging them with dispensing opiates illegally. Within a few years, all the clinics had been closed. Prosecution of private physicians also occurred. Although individual doctors continued to prescribe opiates for addicts (and in some cases served jail terms as a result), leading physicians, with the sanction of the American Medical Association (AMA), came to favor strict enforcement of the Harrison Act. Community-based and private practice options for treatment of opiate addiction had all but disappeared by 1930. The AMA's alignment with these policy initiatives, which impinged directly on the practitioner's freedom to practice medicine without lay interference, reflects the deeply problematic nature of opiates for physicians.

AMA Actions and Physician Attitudes

Opiates in general, and morphine in particular, formed an essential part of the physician's therapeutic armamentarium throughout the period under consideration. At the same time, in the context of Progressive Era reform and the scientific reform of medicine, opiates also symbolized problems of concern to individual physicians and to the American Medical Association, which sought to represent those physicians' interests. Concern about exaggerated therapeutic claims, deceptive labeling, and wholesale swallowing of frequently harmful drugs in America's unregulated medical marketplace led Progressive Era muckrakers like Samuel Hopkins Adams to press for the passage of the Pure Food and Drug Act of 1906. When the Harrison Act came before Congress, although it had originated in diplomatic circles, medical reformers also endorsed its provisions to restrict the sale of opiates. Among the groups voicing support for the legislation was the AMA.

Since its reorganization in 1901, the AMA had acted vigorously to gain greater control of the medical marketplace as part of its campaign to consolidate orthodox medicine's monopoly on legitimate medical practice. In 1906, it created a Council on Pharmacy and Chemistry, which systematically tested drugs for purity and the validity of the therapeutic claims made on their behalf.[50] Through the work of this council and through exposés of nostrums in the *Journal of the American Medical Association,* and in a series of volumes entitled *Nostrums and Quackery,* the AMA discredited many available preparations.[51] It also sought to instill in the American public a narrower sense of therapeutic effectiveness than had prevailed in the nineteenth century. While traditional therapeutics had promised systemic redressment of imbalances in the body, a new view of disease causation ushered in by the bacteriological discoveries of Pasteur and Koch had led to a new view of therapeutic specificity. If diseases were now understood to have specific causes and precise sites of pathology, effective drugs were similarly portrayed as having specific effectiveness against individual ills rather than acting broadly against whole categories of illness.

An important component of physicians' anxieties about medication use and misuse was the widespread belief, within and without the profession, that the chief cause of the prevalence of drug addiction was the medical overuse of opiates. Passage of the Harrison Act meant that patients could no longer obtain opiates on their own for self-medication. However, excessive administration and prescription of opiates by physicians remained a significant hazard, which the AMA sought to eliminate by educating physicians as to the minimum indications for use of opiates. Textbooks of therapeutics and materia medica from 1890 to 1940 reflect a growing concern about the addiction hazard associated with opiates and a progressive narrowing of their therapeutic indications.[52]

When, in 1919, the Supreme Court interpreted the Harrison Act in a manner that in effect restricted physicians' autonomy in making prescribing decisions (by declaring addiction maintenance a violation of the law), the AMA did not object. While it jealously monitored enforcement of the Harrison Act to prevent undue harassment of physicians on minor technicalities, it substantially agreed with the attitude of the court. In both 1919 and 1924, the AMA adopted resolutions opposing ambulatory treatment of addicts—and thus, in effect, addiction maintenance. The AMA also exercised vigilance in communicating with federal and state enforcement authorities and state medical societies to prevent physicians convicted for violations of the Harrison Act from resuming medical practice.

On the Scientific Front: Research and Theories

While policy makers grappled with the problems of enforcement and practicing physicians adjusted to the new environment created by the Harrison Act, changes also occurred in the theoretical debate about the nature of addiction. A brief summary of their research provides the background for observing the crystallization of a subset of these ideas into an official model of addiction in the 1920s and 1930s. Study of the effects of opiates on body tissue followed the general lines of physiological research in this period, and attempts to explain addiction as a disease followed the dramatic changes in understanding of etiology and transmission implied by bacteriological models of disease causation.[53]

In the last quarter of the nineteenth century, as physiological investigations focused on such issues as cell function and the sympathetic nervous system, various investigators devised laboratory experiments to elucidate the effects of continued opiate administration on physiological function. From the 1870s on, physiologists and pharmacologists conducted systematic animal tests to determine the effects of acute and chronic morphine administration. Attempts to correlate methods of administration, doses, and durations of action to pathological findings in autopsy (in human patients and in animals used in research) involved a wide variety of morphological, neurological, and histological examinations. Although various claims were made regarding the explanatory value of these findings, the results frequently proved contradictory. Gradually, a consensus emerged that morphine itself, even when chronically administered over a long period, had little deteriorating effect on body organs or tissues, and that apparent findings to the contrary were explainable by such secondary conditions as malnutrition. By about 1920, these points were generally agreed on.[54]

Others attempted to elucidate the action of opium or morphine in chronic administration leading to habituation. The phenomena of tolerance (the need for increasing doses to achieve the same effect) and dependence (the need to continue taking the drug to stave off withdrawal symptoms) had been described as early as the eighteenth century; but at that time, no unifying concept of addiction had been proposed, and few warned against problems associated with chronic opiate use.[55] By the late nineteenth century, dependence on opiates and on alcohol were lumped together in the general psychiatric category of "inebriety." In the absence of a dominant theoretical explanation for this phenomenon, a diversity of treatment approaches and theoretical explanations flourished.

The key question was whether addiction was a disease or a vice. In 1918, the Treasury Department surveyed health officials regarding views on the nature of addiction. In the responses, 425 health officials stated that the physicians in their communities considered addiction a disease, while 542 reported that they regarded it as a vice.[56] While these results are hardly conclusive, because they simply report health officials' beliefs and fail to make any distinctions within individual communities, they do suggest a lack of consensus among professionals. Those who contended that addiction was a disease sought to demonstrate that it was "a distinct, definite physiological disease condition with definite uniform manifestations and phenomena and a definite understandable causation. . . .The signs and symptoms are as constant, uniform and recurring as those of any other disease."[57] Proponents of the vice theory argued that chronic self-administration of opiates was simply a moral lapse, and that the appropriate response was punitive rather than medical. Others argued that addiction fell somewhere between a vice and a disease, holding that reinforced by its pleasurable effects, drug taking became a habit, and that functional changes in physiology resulted from chronic administration.[58] In 1922, the Committee on Narcotic Drugs of the Medical Society of New York grouped these ideas in a report, saying: "Your committee does not consider drug addiction as a disease entity, but rather as a habit. . . . Functional disturbances of the internal organs follow acute excesses or prolonged use of these drugs. These conditions can be cured by cutting off the drug, hence a relapse to the former habit when opportunity offers."[59] For those who held to the disease theory, the next question was, what kind of disease did addiction represent? The proposed models rested not only on the scientific observations of their proponents but also on the observer's fundamental attitude toward addicts as human beings. As might be expected, explanations based on degeneracy theory or other versions of psychopathology took a generally harsh and punitive view toward addicts; in addition, they held out little hope of cure.

The work of Ernest S. Bishop exemplifies the union between physiologically grounded theories of addiction and a sympathetic view of addicts. Bishop was a vocal proponent of the so-called autoimmune theory of addiction.[60] This idea reflects the power of immunological ideas in this period, as it implicitly treated the drug as analogous to an invading pathogen or the toxin it produced. It posited a response similar to the development of diphtheria antitoxin or antibodies against infectious bacterial agents. This model explained tolerance by theorizing that the body developed an antitoxin that protected it from the dangerous effects of steadily increasing doses of morphine. When morphine administration was stopped, the endogenous substance then exerted toxic effects on various organs, and these effects were observed in the withdrawal syndrome. Bishop's ideas were based purely on clinical observation, but they were bolstered by theoretical underpinnings characteristic of the scientific elucidation of disease in this period. Bishop claimed a mathematical precision in its manifestations: "The narcotic drug administered to an addict suffering withdrawal agonies, will relieve those agonies exactly in proportion to the amounts of drug administered. . . .This is almost mathematical in its working, and an addict after a few trials can tell within a very close margin just how much morphine has been administered by the extent to which it relieves his withdrawal signs."[61] Bishop explicitly rejected any psychiatric explanation for addiction, and, in contrast to the punitive tone taken by some writers, he dwelt on the suffering of addicts undergoing withdrawal.

Joseph C. Doane, M.D., articulated the view that the pathology of addiction lay in defects of character. Doane had joined the staff of Philadelphia General Hospital as chief resident physician in 1914, just in time to witness the local impact of implementation of the Harrison Act. By 1926, when the bridge tender James Martin came to Philadelphia General's narcotics ward for treatment, Doane had been medical director and superintendent of the hospital since 1920.[62]

Philadelphia General, which had long had a ward for alcoholic patients, opened its narcotics ward the first day of the implementation of the Harrison Act, because many addicts were expected to seek treatment for their habits once access to opiates was restricted. News coverage of the ward's opening linked the banning of opiates to the contemporaneous campaign to prohibit prostitution; according to an unnamed reporter, the Harrison Act was "already nicknamed the twin bill to the white slave law." Among the first patients was a man nicknamed "Doc Hypo" because he had "sold 'shots' for 'two bits' for years."[63]

The narcotics ward at Philadelphia General offered the Lambert treatment method, named after the prominent physician, medical reformer, and future AMA president Alexander Lambert.[64] Lambert had become interested in the problem of treating opiate addiction and had joined forces with Charles Towns to advocate a variant of a common treatment method in which hyoscine or scopolamine was administered to ease the patient through withdrawal and powerful cathartics were given for the constipation that was a prominent side effect of opiate addiction.

The Lambert-Towns alliance beautifully symbolizes the melange of old and new, of orthodox medicine and shady practice, that surrounded the treatment of opiate addiction in the 1910s.[65] Failure to reveal the contents of a medication was a central target of ongoing AMA activity to regulate the sale of medicines along lines consistent with professional authority and scientific method, a campaign carried out through the exposure of false claims in *Nostrums and Quackery* and the work of the AMA's Council on Chemistry and Pharmacy. Physicians used their specialized knowledge and licensing procedures, not secret formulas, to buttress their claims to competence, and the scientific research that would be the source of new medicines depended on open publication of results.

Towns originally trumpeted a secret formula that cured addiction, but his secret ingredients, when revealed, proved to include medicines with modern, scientific panache—hyoscine and scopolamine. Scopolamine was just then in vogue as an obstetric anesthetic that provided a greater margin of safety than either ether or chloroform, which had been used since the 1840s to anesthetize women in labor and childbirth. In exercising its anesthetic effect, scopolamine induced a delirium, followed by partial amnesia. The drug required close monitoring of patients, who might thrash about while delirious and cause themselves injury. Scopolamine appealed to affluent, middle-class women, who both saw it as safer and valued a few days of recuperation from childbirth in a hospital as a respite from their domestic responsibilities. Hospitals, undergoing transition in this period from charitable institutions to scientific ones, welcomed a drug that made childbirth an event best managed in the hospital.[66]

The drug's somewhat lurid side effects, including bizarre delusions, intense flushing, and wide pupil dilation, were nevertheless consistent with a long therapeutic tradition in which powerful and visible drug effects were seen as evidence of efficacy. At the same time, hyoscine and scopolamine were going out of favor in mental asylums as drugs to control patients' disturbed behavior.[67] From this perspective, it appears that Towns's patients were getting obsolete treatments borrowed from asylums for the insane. Lambert, within a few years, disassociated himself from Towns, but the fundamentals of the treatment regimen they championed were widely used in hospitals treating addiction into the 1920s, including at Philadelphia General.

The demand for treatment in Philadelphia General's newly opened narcotics ward overwhelmed all expectations. By the end of its first month of operation, several hundred patients had received treatment. Three men's wards accommodated up to 117 patients at a time, and a women's ward had beds for up to 40. Considerable optimism prevailed in these early weeks, as physicians, politicians, and the news media believed that, once a round of treatment had gotten patients through withdrawal to a state of abstinence, effective enforcement of the Harrison Act would make opiates unavailable to addicts and relapse impossible. As chief resident, Joseph Doane oversaw the treatment of addicts with a complex regimen of medications (which, in addition to the hyoscine and cathartics, included a quick taper off of morphine, and such adjuncts as barbiturates to aid in sleep and small doses of strychnine as a stimulant). The ward physicians also exhorted patients to recognize that, after their release, they would not be able to acquire new supplies of opiates, and that they must develop sufficient willpower to face the rest of their lives without drugs.[68]

By 1920, Doane was considerably less sanguine about the prospects for his addicted patients. He challenged the view that toxemia was sufficient as an explanation of addiction, as Bishop and others had claimed. Rather, Doane argued, a neurological taint was the critical etiological factor. It was precisely this taint that separated the addict who relapsed repeatedly from the addict who was cured the first time he was detoxified. Doane noted the difficulty in diagnosing addiction. Addicts appeared normal when taking their drug: "A diagnosis can only be certainly made after a reasonable period of isolation, with no medication."[69] In other words, it was frequently necessary to induce the withdrawal syndrome to clinch the diagnosis. Addiction was thus an ambiguous condition that could be hidden even under close examination by a physician as long as the addict had reliable access to drugs, but that quickly became manifest when the well-recognized withdrawal syndrome set in.

Psychological defects were both character-defining and difficult to detect: this formulation became central to the view of addiction that soon dominated medical and policy views, most influentially through the work of Lawrence Kolb (discussed in Chapter 5). In the meantime, the Rockefeller-funded Bureau of Social Hygiene, initially created to study prostitution scientifically, formed a committee to study the phenomenon of drug addiction, which seemed so prevalent in the social world frequented by prostitutes.

2

The Opportunistic Approach

New York was the first American city to appoint a vice commission to study the problem of prostitution. Its Committee of Fifteen was established in 1900 and produced its report, *The Social Evil,* in 1902. Across the United States, social hygienists and social purists publicized what they saw as the evils of prostitution, and they persuaded Congress to appoint a commission to examine the national scope of the problem in 1907. This body's conclusion that the "white slave traffic" was a real and serious phenomenon fueled a new round of local energies across the nation.[1] In New York City, a grand jury was appointed in 1910 to examine the white slavery problem, and John D. Rockefeller Jr. served as its foreman. This experience prompted Rockefeller to urge systematic and scientific investigation of prostitution and the closely related problem of venereal diseases, especially syphilis. As steward of the funds his father had made available for benefaction, he was well placed to act on his interests in reform, and in 1911, with the banker Paul Warburg and the lawyer Starr Murphy, he created the Bureau of Social Hygiene to that end. The bureau hired George Kneeland, director of the team of investigators that had produced *The Social Evil in Chicago,* to develop a comparable report on prostitution in New York, *Commercialized Prostitution in New York.*[2]

The Bureau of Social Hygiene's investigations identified drug use as a problem closely linked to prostitution, and, in 1919, it created a Committee on Drug Addictions to study the drug problem scientifically. Allan Brandt has characterized the bureau's work on prostitution as representing a shift from moralistic purity crusades to a reform that was "efficient, scientific, elitist."[3] Ellen Fitzpatrick has described it as exemplifying the transition from amateur to professional and scientific reform that characterized this period.[4] I argue, with Adele Clarke, that the bureau also played an important, if transitional, role in transforming the nature of scientific research into the issues of deviant sexuality and drug use that it studied.[5] With respect to both sex and drugs, the bureau surveyed prominent scientists regarding the most promising research approaches to these social problems, supported some research in an opportunistic fashion, and proffered its resources to the National Research Council, which coordinated narrowly focused scientific projects, rather than wide-ranging investigations by independent researchers.

The Bureau of Social Hygiene accordingly played a transitional role in several respects. Having identified important social problems involving deviant behavior that had been brought to public attention by social reformers, it invited laboratory scientists, elite clinicians, and, to a lesser extent, other experts and a few leaders in social reform, to propose what forms of research might best improve understanding or control of these problems. In so doing, it promoted the professionalization of reform. Through its formation of expert committees and extensive surveys of elite academic and professional opinion, it stimulated the formation of networks of researchers interested in related problems and helped develop consensus on the status of current research and, to a lesser extent, on what research approaches would be most useful. It thus performed a clearinghouse function that made it easier for entrepreneurial scientists to perceive emerging hot research areas and capture them for their own laboratories. In this way, it contributed to the transformation of science patronage of the 1920s, especially in framing scientific research as relevant to social problems.[6] The bureau stopped short, however, of setting out a coherent research strategy of its own, in part because none emerged clearly from its surveys of scientific opinion. Guided by a naive faith in "excellence," which it implicitly equated with elite social and academic status, it embraced an opportunistic approach that called for funding those projects brought to its attention that appeared worthy.[7]

Over the course of the 1920s, Bureau of Social Hygiene's leadership increasingly framed drug addiction as a problem of criminology rather than medicine. In so doing, it became increasingly aligned with the federal drug policy infrastructure based in Washington, D.C. The bureau's Committee on Drug Addictions also had explicit connections to the drug policy arena on both the domestic and international fronts; its two directors both had backgrounds in law enforcement, and its members included a former New York City police commissioner.

Similarly, the Committee on Drug Addictions linked its work with the ongoing international effort to control worldwide supplies of opiates, centered since 1919 at the League of Nations headquarters in Geneva. Nations participating in this process agreed to determine their level of medical need for opiates and restrict imports to those levels. This plan pertained to countries where opium poppies were not grown, and that therefore relied on imports for their supplies of opiates. Opium-producing countries were expected to cooperate by exporting only in conformity with quotas set by this assessment of needs. This regime was expected to exert effective control over global supplies of opiates, because at that time, virtually all processing of raw opium into morphine, heroin, and other products occurred in pharmaceutical plants in a few European countries; locating processing plants near growing sites, as a means of evading legal detection of drugs, had not yet occurred. The United States participated in this process despite its refusal to join the League of Nations.[8]

The Bureau of Social Hygiene conducted surveys in seven American cities to determine the amounts of opiates and cocaine used in accordance with the Harrison Narcotic Act, and the results were forwarded to the Opium Advisory Committee of the League of Nations to help determine an import level to meet the legitimate medical needs of the United States (including strategic stockpiles for use in case of unexpected shortages or in times of war). This effort was predicated on the desirability of drug prohibition as a policy to reduce opiate addiction to a minimum.

With respect to the social problems of prostitution and drug addiction, all of the bureau's activities accompanied and were consistent with a shift from the optimism, meliorism, and social and environmental explanations of the early Progressive Era to the focus on defective individuals, deployment of diagnosis and triage, and reliance on laboratory research that characterized the 1920s. But despite this trend toward a punitive model, the Committee on Drug Addictions was also home to a dissenting voice in the national policy debate. To oversee its work, the bureau hired Dr. Charles E. Terry, who had come to regard drug addiction as a public health problem in Jacksonville, Florida, where as city health officer he had opened the first clinic to register addicts and provide them with maintenance doses of morphine. As executive officer of the Committee on Drug Addictions, Terry conducted surveys of medicinal opiate use in U.S. cities and compiled information about all aspects of addiction. His findings were published in 1928 in a volume entitled *The Opium Problem.* Terry's case illustrates the difficulties confronting a physician convinced that addiction was a disease, whose victims deserved humane treatment, at a time when the demographics of addiction were shifting from iatrogenically addicted people in all walks of life to a narrower demographic segment centered in the world of urban amusement and vice.[9]

The Committee on Drug Addictions

On January 11 and 12, 1919, several officials of the Rockefeller philanthropies met with the eminent Johns Hopkins professor of medicine William H. Welch, the renowned public health expert Hermann M. Biggs, and Simon Flexner, director of the Rockefeller Institute for Medical Research, to discuss a proposal put before them by Arthur D. Greenfield, a New York attorney.[10] Greenfield's experience working on the New York draft board had interested him in the problem of opiate addiction. A vocal opponent of maintenance, Greenfield sat on a number of expert committees on narcotics; with Emil J. Pellini, he also co-authored a study refuting the theory that addiction resulted from the body's production of protective antitoxins against morphine. He may have owned one or more treatment sanitaria.[11]

Greenfield suggested that the Rockefeller Foundation create an institute for research into the causes and cure of drug addiction and for treating patients. The proposal's backers included Surgeon General Rupert Blue, the addiction expert Dr. Ernest Bishop, and Charles Terry. Greenfield cited estimates of between 20,000 and 500,000 addicts in New York City alone (although mentioned by various observers, these figures were significant exaggerations).[12] While some medical experts argued that addiction was a disease that could be understood in physiological terms, explanations like the autoimmune theory advanced by Bishop and others lacked satisfactory proof. Individuals without the means to pay for a stay in a private sanitarium, Greenfield noted, had virtually nowhere to turn for help. He suggested that both private and public bodies were awaiting scientific leadership before tackling the problem.

The group agreed that creating such an institute would run counter to the Rockefeller Foundation's policy in several respects. The foundation typically helped others start projects and then withdrew from participation. In the area of addiction, it would be impossible to steer entirely clear of the complex political issues involved, especially those surrounding control of drug trafficking. Nor did the foundation engage in direct delivery of services like treatment. The consensus was that some initial study would be the best course, and that the research questions should be referred to the Rockefeller Institute. However, the institute's director, Flexner, saw the problem as chiefly relating to social life and personal habits; his views may explain its failure to take up the problem.

In 1921, Raymond B. Fosdick communicated to John D. Rockefeller Jr. the results of several meetings called to discuss the problem of drug addiction, whose dimensions were said to be reaching alarming proportions.[13] Those present at these discussions included Charles Terry, former New York City Police Commissioner Arthur Woods,[14] and Dr. Thomas Salmon, a leading psychiatric reformer. Fosdick noted that, while various disciplines had studied the problem of addiction, no systematic attempt had been made to assemble all the relevant data and, on that basis, to create a positive and coherent program of action. The possibilities discussed included asking the National Research Council (NRC) to undertake a study, but it was decided that the problem needed to be approached from many different angles, and the NRC, as then organized, was judged not capable of mounting a satisfactory effort. The group also discussed involving the American Social Hygiene Association and the Mental Hygiene Association (reflecting, respectively, the perceptions that drug addiction was connected to the problem of prostitution and that it represented a form of mental illness).

Fosdick and the others concluded that the best plan would be to have the Bureau of Social Hygiene undertake one year's work as a first step. The group suggested that Rockefeller allocate $12,000 for the creation of a committee, whose membership would include Katharine Bement Davis, director of the Bureau of Social Hygiene, Salmon, Dr. Snow of the American Social Hygiene Association, Woods, and Fosdick. Davis, a prominent figure in prison reform who had served as the chair of the New York City Parole Commission at its inception, was named chair of the committee.[15] The funds would be sufficient to hire one investigator (Terry assumed this position) and two assistants. The intention, as in the bureau's work in prostitution, was to lay a scientific foundation to guide future policy. The Bureau of Social Hygiene's Committee on Drug Addictions was thus created.

By early 1924, the committee had agreed on a three-pronged program, to include education efforts, sociological research, and "pure research."[16] The educational work was specifically aimed at physicians, reflecting the view that misprescribing of opiates remained an important cause of addiction. Sociological research was broadly seen as covering drug trafficking and the economic aspects of addiction. Pure research was expected to involve a wide range of approaches. Medical concerns were again prominent: the legitimate need for the drugs must be assessed. This effort would be linked to international efforts to control world supplies of opium and its derivatives. After World War I, the League of Nations assumed leadership of the activities launched at conferences in Shanghai in 1909 and The Hague in 1911 to regulate the global opium trade. Beginning with an agreement between the governments of China and India in 1907, international efforts focused on control of opium supplies as a means of limiting the incidence of addiction. Study of the causes of addiction or other efforts intended to reduce the demand for drugs were consistently rejected.[17]

Psychological research should determine the differences between addicts of high social class and underworld addicts; it should also examine the relation of addiction to personality type, psychoneuroses, and so forth. Desirable laboratory research tasks included testing the autoimmune theory advanced by Bishop and others, studying metabolic pathways of opiates in the body, and seeking substitutes for opiates in medical practice. The committee envisioned itself as a central coordinating body giving support to and conferring coherence on work largely carried out by other groups—for example, the educational work by the National Health Council, the American Social Hygiene Association for the sociological research, and the National Research Council for the laboratory studies. At this stage, the Bureau of Social Hygiene allocated $50,000 to launch these efforts and to continue a survey already begun of the medical and scientific literature on addiction.

While discussion about funding work to be carried out by such groups as the NRC continued without definitive action, the pattern of the committee's own activities had emerged by the mid 1920s. On the sociological side, Charles Terry directed a series of surveys of physicians and pharmacists in six American cities to determine the precise medical need for opiates. Apparent variation in per capita consumption among different cities suggested the need for further sociological investigation to determine its causes, and Terry later undertook a survey with refined methods in a larger city (Detroit) to provide a better basis for extrapolating to national levels of the need for medicinal opiates.

These city surveys, while framed as sociological research, were tied to practical objectives. They also reflected the committee's links to a wide range of other organizations and institutions embarked on the combination of social and scientific reform that characterized the period. The main purpose of surveying actual and medically required levels of opiate use was to lay the basis for estimating the amount of opiates needed for medical practice in the whole country. Since opiates were imports (originating as raw opium in poppy fields in southern Asia and the Middle East and passing thorough pharmaceutical factories in Europe and, to a lesser extent, in the United States), limiting the amount entering the country to what was required for medical practice seemed an appropriate goal for controlling use. Moreover, the Opium Advisory Committee of the League of Nations had persuaded the State Department, despite the U.S. refusal to join the League, to take part in the creation of an international regime to control opium production; determining America's legitimate medical need for morphine and codeine was essential to this effort. The Committee on Drug Addictions made its survey results available to the Public Health Service, which was attempting to determine the level of national need for opiates at the behest of the State Department. It also reported its findings to the annual drug conferences of the League of Nations at Geneva, where representatives of various countries continued the effort to achieve worldwide control of the opiate market that had been the original impetus for passage of the Harrison Act.

In addition to working on the city surveys, Charles Terry, with the assistance of Mildred Pellens (who, while working for the committee, also pursued a medical degree) continued the literature survey of scientific opinion on addiction. The pure research aims of the committee were somewhat narrowed following the addition of two medical scientists from elite universities to the committee: the famous Harvard bacteriologist Dr. Hans Zinsser urged a range of psychiatric and physiological studies, while Dr. Lafayette Mendel, professor of biochemistry at Yale, recommended study of the metabolic fate of morphine in the body.[18] The committee provided funds to the eminent University of Pennsylvania pharmacologist Alfred Newton Richards for such studies, and, on Richards's recommendation, supported Dr. O. H. Plant, professor of pharmacology at the State University of Iowa.[19] Plant's laboratory had already studied the physiological effects of morphine, but the work had languished for lack of funds. Plant wanted to hire a chemist to aid in the development of improved methods for detecting morphine in various body tissues. Mendel's recommended research direction thus resulted in the committee's selection of physiological studies of morphine effects from among the range of possibilities the committee had envisaged earlier.[20]

Soon thereafter, Richards joined the committee as a member and added his call for greater coordination of the scattered avenues of research on opiate addiction.[21] However, the work at his laboratory—studies of the distribution of morphine in the body and the pathways of its metabolism and excretion—were standard tasks in the development and assessment of medicines and held little promise of elucidating the nature of addiction. The committee's continued support of this work reflected not only Richards's stature and value as a committee member but the ways in which morphine's medical status remained central to committee concerns.[22]

Richards's value to the committee was evidenced when he put it in touch with Dr. Joseph Doane, who oversaw the clinical ward for opiate addicts at Philadelphia General Hospital. In November 1925, Charles Terry traveled to Philadelphia, where he judged Doane's study of his addicted patients, which included sociological, physiological, psychological, and pathological studies, the most thorough clinical research on addiction yet undertaken. Under Committee on Drug Addictions auspices, and with funding to carry out three years of physiological and psychological study of the clinical material available on Philadelphia General's narcotics ward, the Philadelphia Committee for Clinical Study of Opium Addiction was formed.[23] Members included Richards, Doane, the psychiatrist Edward Strecker (a prominent figure in the movement to reform psychiatry), and the pharmacologist Horatio C. Wood Jr. The research was done by a team consisting of Dr. Arthur B. Light, Dr. Edward G. Torrance, Dr. Roy B. Richardson (a psychiatrist), a psychologist, a chemist, two nurses, and a secretary.[24] The most important results of this group's exhaustive study of morphine and heroin addicts included the determination that the withdrawal syndrome following abrupt cessation of chronic opiate administration, although hellish to experience, was not life-threatening in the absence of other medical complications. In addition, an exhaustive series of negative findings established that no clear physiological attributes distinguished addicts from nonaddicts.[25]

The Opium Problem and the Clinician's Dilemma

The committee's pattern of activities and research support in the mid 1920s reflected the Rockefeller philanthropic methods of bettering the condition of mankind through scientifically informed research. Its universalistic aspirations—the search for a definitive solution to a poorly understood social problem by tackling research from every perceived angle—also reflected a positivist notion that once sufficient facts were gathered, a solution would emerge. The committee's acknowledgment that not enough was known about addiction led it to cast a wide net and to consult a broad range of experts in different fields rather than to develop a tightly focused project based on a unitary idea of what the problem was. The individual research directions chosen reflected shifting committee membership, links to other organizations, and a general mission of supporting the creation of knowledge rather than providing direct services.

The committee's most enduring accomplishment was the publication of Charles Terry and Mildred Pellens's *The Opium Problem,* the result of their exhaustive research in the European and American medical and scientific literature on opiates, as well as of Terry's experience in Jacksonville and New York.[26] This work clearly framed addiction as a disease and argued forcefully that implementation of the Harrison Act, and especially the ban on maintenance and the closing of the municipal clinics, had worsened the plight of addicts.[27] Like many observers, Terry and Pellens noted that the ban on nonmedical sale of opiates stimulated the formation of a widespread, organized illicit market. But Terry was perhaps the best-positioned person in America to appreciate the impact of these policies on the medical treatment of addicts. Given addiction's long-standing status as a stigmatized condition, he and Pellens noted, many physicians already had little interest in it. Medical schools failed to include addiction treatment in their curricula (a situation that would hardly improve over the subsequent decades).

Treasury Department actions had only made this situation worse. Implementation of the Harrison Act had created huge new bodies of patients seeking medical help for addiction, and the opportunity to engage these patients in a meaningful process of study and treatment had been lost. The first impact of the Harrison Act was to end the unregulated sale of opiates in pharmacies, and many addicts who had maintained their habits from this source initially sought help from physicians. This trend, Terry and Pellens believed, could have meant proper medical treatment for thousands of addicts who had not sought help before. However, as the Treasury Department moved against physicians and pharmacists deemed to be prescribing or dispensing too many opiates, both professions reacted with fear. Any physician who accepted large numbers of addicts as patients and began them on a program of maintenance, a logical step to take between 1915, when the Harrison Act was first implemented, and 1919, when the Supreme Court ban on maintenance was handed down, would inevitably come to the notice of Treasury officials monitoring amounts of opiates prescribed by individual physicians, as would any pharmacist filling substantial numbers of such prescriptions. Questions about this practice led to the Supreme Court cases that culminated in the 1919 decisions. As Terry and Pellens noted, even if an indictment brought by the Treasury Department did not result in conviction, the prospect of the publicity alone deterred many physicians and pharmacists from serving addicts. Faced with the retreat of physicians and pharmacists, Terry and Pellens stated, "patients wandered from one institution to another, bought this remedy and that, applied in vain to their family doctor for suggestions or help, and, finally, impoverished, and discouraged, gave up the fight and as best they could continued a miserable existence the reasons for which they could not understand and for which, in many instances, they were not responsible."[28]

The opening of the municipal narcotic clinics had prompted another wave of patients, in particular indigent ones, to seek treatment for their addiction and had been another opportunity to keep them engaged in treatment; but enforcement of the Harrison Act "resulted in their closing before their development could bear fruit through practical administrative and medical procedures."[29] This language, reflecting Terry's own experience in managing a maintenance clinic in Jacksonville and his observation of mismanagement at New York's Worth Street clinic, undoubtedly reflected his faith that, with longer experience, maintenance could be developed into a well-understood and effective means of managing addiction and reducing addicts' suffering. For Terry and Pellens, whatever the merits of banning the nonmedical sale of opiates, the Treasury Department's enforcement policies, especially the overzealous punishment of physicians and pharmacists and the closing of the clinics, represented an enormous lost opportunity to learn more about treating addiction from working with the large numbers of addicts whose addiction was forced into the open by the ban on nonmedical sales.[30]

Terry and Pellens had compiled a bibliography of over 4,000 items in developing *The Opium Problem*. In 1926, once they had whittled an unwieldy preliminary version to manageable, although still exhaustive, length, the committee discussed whether to publish it and, if so, in what form.[31] Despite a generally favorable reception of the manuscript, George McCoy, director of the Public Health Service's Hygienic Laboratory, objected to publication. He had consistently disagreed with Terry's views on the nature of the addict population. In a 1924 discussion of whether the committee should provide guidance to medical schools in educating about addiction, McCoy had asserted that such education was unnecessary, because the problem of people addicted through medical treatment had virtually disappeared.[32] The implication was that the current crop of addicts were denizens of the underworld; there was no need to educate physicians about a problem confined to such types. Terry argued the opposite view, believing that many addicts still resulted from medical mismanagement. As Terry knew through his long experience treating addiction, many opiate addicts ascribed their use of morphine or heroin to both medical and recreational reasons; others provided reasons that could not easily be classified as one or the other.

This issue goes to the heart of the dilemma faced by physicians like Terry who advocated humane medical treatment of addicts in the 1920s. Like virtually all observers, he made a crucial distinction between a person who became addicted to opiates through treatment of a medical condition and a person who combined drug use with behavior clearly considered deviant. This distinction paralleled the distinction between innocent and culpable sufferers from syphilis that pervaded Progressive Era reformers' socioepidemiological understanding of the disease: wives, and the infants born to them, were innocent of any moral taint, while prostitutes, the putative source of the disease, were cast as guilty. Straying husbands or bachelors sowing wild oats held an ambiguous status between guilt and innocence: their behavior in resorting to prostitutes was certainly blameworthy, but campaigns to eradicate the double standard and to instill the idea that careless casting of a man's seed was detrimental to his health implied that the man was reclaimable through education.[33] Similarly, commentors on the addiction problem repeatedly distinguished innocent medical addicts from guilty ones who contracted the habit through vice.

Terry's commitment to maintenance was based on his belief that most addicts were normal individuals who had become mired in addiction, a true, physiological disease condition, through circumstances that were at least to some extent beyond their control. He sympathized with addicts who were driven to the illicit market when legal supplies of their drug dried up and argued that enforcement of the Harrison Act increased the likelihood that addicts would steal to support their habits.[34] However, he was less prepared to defend the truly deviant addict. Like other physicians urging humane treatment for addicts in the 1920s, he hit an ethical blind spot when confronted with "vicious" addicts who were also criminals.

Yet many addicts fell into a more complex or ambiguous category between these two extremes. And when they did become visible to researchers, as at Philadelphia General Hospital, they were often far along in addiction and experienced in a repertoire of habit management behaviors, including adeptness at navigating the black market or, perhaps, at wheedling opiates from physicians. Understandably, many physicians did not embrace such individuals as desirable patients. Even undesirable patients have an ethical claim on a physician's competent and compassionate care, but Terry, and other physicians who advocated maintenance, such as E. H. Williams, were bucking a trend that culminated in apparent medical justification for general practitioners to reject addicts as patients.[35] Contributing to it were the federal, AMA-endorsed ban on ambulatory treatment of addicts, the lack of any consensus on effective treatment, and the mounting therapeutic pessimism that resulted from growing recognition of addiction's intractability in the face of available treatment resources. As David Courtwright has said, Terry's point of view was defeated, above all, because few Americans were prepared to sympathize with the type of addict who predominated by the late 1920s.[36] But to a certain extent, Terry implicitly conceded the battle by basing his case on the argument that the preponderance of addicts were ordinary people caught in something beyond their control. If addicts' claims to humane medical care were to be based on their appeal as patients, those whose behavior was deemed undesirable were implicitly excluded. Doctors' distaste for addicted patients in this period is understandable, but their collusion in denying addicts medical care violated a fundamental ethical value underlying the extraordinary cultural authority medicine acquired in the twentieth century: that the physician's obligation to the individual patient overrode all other concerns. The medical retreat from addicts as patients helped make possible the norm of incarcerating addicts for their condition, a fate that became increasingly common in the 1920s and 1930s.

Terry won a temporary victory, however, and McCoy resigned from the committee over his displeasure with the manuscript. This dispute had an additional effect of stimulating the committee to conduct a broad survey of scientific and medical opinion on addiction as an adjunct to a poll of experts around the nation on the desirability of publishing *The Opium Problem*. The committee instructed Pellens to develop a questionnaire whose immediate objective was to seek resolution of the dispute over whether to publish *The Opium Problem*, and, if so, in what form, and aimed at what audience. Additional questions, however, polled respondents on what directions future research on addiction should take. Guidance was necessary, because Terry and Pellens' literature survey had revealed a complete lack of consensus on such issues as the etiology of addiction, its extent, its fundamental nature, or the best methods of treatment or control.[37]

A total of 122 individuals were polled, including 82 physicians and scientists (pharmacologists, physiologists, biochemists, internists, etc.), 28 psychologists and psychiatrists, and 12 sociologists. The former two lists consisted of recognized leaders in the medical sciences, but the "sociologists" were a mixed bag of academic sociologists and heads of charitable reform organizations interested in the drug addiction problem. The survey method consisted of visiting each individual (where feasible; otherwise, communication was by mail), acquainting him with the concerns of the committee, and giving him a copy of Terry and Pellens's manuscript.

During a second interview, the questionnaire was administered. The tabulated results showed that most of the 77 respondents believed that research would be useful in most of the areas named in the questionnaire; these broke down into physiological, psychological, and sociological topics, paralleling the disciplines of those polled. Where individuals' specific suggestions were recorded, a lack of clear programmatic vision for addiction research is similarly evident. Most made vague suggestions in general areas. By far the largest number suggested physiological research on the effects of morphine in the body and on specific mechanisms of tolerance and withdrawal. While many responded affirmatively regarding the desirability of psychological or psychiatric research, few had specific suggestions in this area. Five linked research suggestions to drug legislation and control; four mentioned eugenics or heredity; and ten recommended research on the effectiveness of treatment.

Where suggestions were made, they either covered too wide a range of possibilities to reflect a clear recommendation or were narrowly stated projects that fell within the respondent's own field, with no clear indication of what this research would contribute to a more effective understanding of addiction. A few are worthy of specific mention. Robert Yerkes, the famous Yale primatologist and chairman of the Bureau of Social Hygiene's Committee on Research on the Problems of Sex, suggested that the similarity of primates to humans made them excellent candidates for psychological and social, as well as physiological, studies of addiction; his recommendations were supported by Simon Flexner. The Harvard pharmacologist Reid Hunt, whose later involvement with the committee's successor body at the NRC is examined in Chapter 3, tersely recommended development of better medications to replace morphine. Lawrence Kolb, a Public Health Service psychiatrist, whose influential model of a psychiatric etiology of addiction is discussed in Chapter 5, believed his own work in this area had largely obviated the need for further psychological investigations; therefore, his recommendations were for more physiological research.

Such a diffuse set of recommendations yielded little guidance in shaping the direction of research. However, endorsements for the publication of *The Opium Problem* were strong enough to persuade the committee to issue the work, which appeared in book form in 1928. Besides containing Terry and Pellens' arguments regarding the treatment of addicts, the work exhaustively surveys the full range of research on opiates and opiate addiction up to that date, including physiological studies in animals and humans, clinical studies, assessments of the scope and nature of addiction as a social problem, and theories of the etiology and mechanism of addiction, with lengthy quotations. Despite little early recognition, this work remains a classic and standard reference today.

In the meantime, the committee was also active behind the scenes in the legislative realm, despite an avowed stance of nonpartisanship. Staff member Lawrence B. Dunham's earlier career had been in law enforcement under former New York Police Commissioner Arthur Woods, who was now acting president of the Board of Directors of the Laura Spelman Rockefeller Memorial, the body that funded the Bureau of Social Hygiene. Woods also worked with the Opium Advisory Committee of the League of Nations, on which he served as U.S. assessor. Dunham worked with this committee as well, communicating regularly with heads of various government bureaus responsible for some aspect of drug control.[38] He advised Prohibition Commissioner Levi Nutt on such matters as proposed amendments to federal drug legislation and setting quotas for imports of raw opium by American pharmaceutical companies. Dunham was not knowledgeable about all aspects of government-sponsored work on addiction, however. His misidentification of Lawrence Kolb as a pharmacologist in 1926, in citing a work on the estimated prevalence of addiction in the United States that Kolb had co-authored, suggests Dunham was not familiar with Kolb's several articles on the psychiatric etiology of addiction published the preceding year.[39] Dunham's focus on enforcement issues and his interest in participating in the international drug control efforts centered in Geneva became more significant when he succeeded Katharine Bement Davis, who was retiring because of illness, as chair of the Committee on Drug Addictions in 1928.

During her tenure as chair, Davis had also concluded that controlling the worldwide opium supply offered the most promising route to control of the problem of addiction. Citing Pellens's questionnaire results, as well as her own reading and her seventeen years' experience in working with addicted criminals, Davis recommended that the committee give this the highest priority. The city surveys of levels of medicinal opiate use provided data in direct support of this program. The educational activities envisioned as most important in 1924 received no mention in Davis's suggested program for 1927.[40] The Bureau of Social Hygiene had also helped fund a commission that investigated the feasibility of reducing the world opium supply by means of a crop-substitution program in Persia.

In fact, Pellens's questionnaire had not revealed significant positive support for a focus on drug control. In a negative sense, though, it did so by failing to suggest a consensus on a single direction that might have guided the committee in developing a coherent research program. As Davis noted, the experts had suggested work in every pertinent field. This meant virtually any proposal merited support, but it also meant there were no compelling grounds for supporting any one in preference to any other. Davis concluded that the committee should continue to support the ongoing physiological investigations of Plant at the University of Iowa and Richards at the University of Pennsylvania. Beyond that, she recommended an opportunistic approach—that is, encourage applications and support the research that seemed most worthy. Thus, the committee did not take a firm stand on specific research directions.

Dunham's view of how scientific research might be deployed against the addiction problem is revealed in his correspondence with Frank P. Underhill, a Yale professor of pharmacology and toxicology.[41] Early in 1927, Dunham read an article by E. Poulsson in the December 1926 issue of *World's Health* indicating that abuse of cocaine had been reduced by the development of newer drugs that had local anesthetic properties but lacked cocaine's undesirable effects. Curious about the possibility of manufacturing improved drugs, Dunham was referred to Underhill by Abraham Flexner.[42] Underhill replied that, while none of the newer local anesthetics was entirely problem-free, several had marked advantages over cocaine and had resulted in reduced use of cocaine in medical practice. Underhill and his associates were, in fact, engaged in trying to develop improved cocaine substitutes, but the work was going slowly because of lack of funds.[43] Dunham did not pick up on this hint with an offer to support the project. More significantly, his failure to follow through on Reid Hunt's morphine substitution idea emphasizes how completely this idea, which became the focus of addiction research funded by the Bureau of Social Hygiene in the 1930s (as discussed in Chapter 3), was imposed on the bureau rather than arising from within it.

At this point, Dunham's mentor, Colonel Arthur Woods, was appointed assessor of the Opium Committee of the League of Nations; his function would be to advise on police matters regarding the illicit drug traffic. Dunham now envisaged another way in which chemical research might further the aims of the Opium Committee. He asked Underhill whether it was feasible to develop a series of chemical tags for individual batches of opiates—additives that, on performance of a simple test, would turn a characteristic color or otherwise identify a particular batch. Since illegal shipments of opiates were typically packaged in unmarked containers, it was impossible to trace them to their point of origin when customs or other officials intercepted them. If each of the approximately two dozen firms known to manufacture morphine and heroin from raw opium were required to incorporate a unique identifier with its shipments, then those firms selling to illicit traders could be traced. Or, if this was not possible, perhaps all manufacturers in each of the half dozen countries that exported medicinal opiates might use the same tag.[44]

Underhill identified ten substances that he believed could be mixed with batches of opiates without adversely affecting their potency or action and that could later be detected with varying degrees of sensitivity.[45] The plan was never carried out, no doubt because of concerns about the potential toxic effects of such additives and the failure of such a procedure to comply with purity regulations governing the manufacture of medications.[46] Dunham, however, was not persuaded by these objections. He noted that risks of toxicity were an accepted component of therapeutics, citing the toxic effects of horse sera used to treat scarlet fever. Furthermore, he cited a medical consensus that incurable addicts were psychopathic cases without a known hope of cure. (In the year since he had misidentified Kolb as a pharmacologist, Dunham had apparently become aware of the growing influence of this explanation of intractable addiction, whether through Kolb's work or through that of Joseph Doane. This idea fitted well with Dunham's own criminological approach to addiction.) The situation was serious enough, in his view, to justify some risk in hope of achieving effective control of the drug market.[47]

Dunham's view of the addiction problem reflected a set of attitudes that increasingly underlay policy and legislation, as well as scientific attitudes toward addiction. As he recognized, the addiction problem was inextricably bound up with advancing medical technology. The pharmacological extraction of pure drugs like morphine from raw plant material and development of even stronger semi-synthetic analogs, such as heroin, now delivered in more potent doses by hypodermic syringe, increased the social risk of addiction. For Dunham, the solution must also arise from the technological matrix that was producing more problematic drugs. Even newer and better drugs might succeed in displacing problematic ones, and failing that solution, improved technological surveillance might increase the possibilities of drug control. The drug-tagging idea reflected Dunham's conclusion that the best point of intervention was at the manufacturing sites in industrialized Europe rather than in countries where the cultivation of drug-producing plants was inextricably interwoven with traditional peasant culture and livelihood. The incurability implied by the psychopathic explanation of addiction further emphasized the desirability of keeping drugs out of the hands of potential addicts.

Conclusion

The Bureau of Social Hygiene's Committee on Drug Addictions did not, in the 1920s, give new shape to research efforts on addiction. The physiological research projects supported by the committee furthered knowledge about morphine metabolism, but significant breakthroughs in the understanding of addiction did not occur. Perhaps in part because of Salmon's early resignation from the committee, psychiatric or psychological research was not directly supported beyond the psychiatric work under Doane at the Philadelphia General Hospital, the results of which the committee found unsatisfactory and did not see fit to publish. The committee's sociological work, based on surveys of physicians and pharmacists, was at best an indirect approach to determining levels of unauthorized use of opiates. The city surveys did not directly seek to examine addicts, their behavior, or their social roles.

The shift in emphases following Dunham's appointment as chair of the committee paralleled hardening policy views toward addicts and deepening pessimism among physicians about the possibilities of effective treatment. Dunham's views as revealed in his correspondence with Underhill illustrate several important points about the mix of science and policy that formed the context for the development of a dominant model of addiction by 1940. Dunham was committed to a supply-side view of addiction control: if worldwide supplies of opiates could be adequately monitored and limited to medical needs, unauthorized use of opiates would be eliminated. Both the development of new drugs as substitutes for opiates and the belief that addicts were psychopathic were consistent with supply-side policy: newer and better medicines would obviate the need for addictive ones, and the explanation of addictive drug use in terms of inherent defects of personality reinforced the idea that the only solution was to keep the problematic drugs out of reach of such individuals. These two elements became the basis for a consensus regarding addiction that dominated medical and scientific views for several decades.

In the committee's diffuse approach to research, the idea of developing nonaddicting substitutes for morphine was but one among many. Drug development as an avenue of approach to the addiction problem had been voiced by only a small minority of Pellens's questionnaire respondents, in contrast to the large numbers who had urged physiological or psychological research. Dunham's own expressed view regarding the possibilities of drug development was confined to discussion of cocaine substitutes whose existence was already known. He lacked sufficient knowledge of the process of drug development to envisage a focused search for similar substitutes for opiates. His interest in the technological possibilities for achieving drug control quickly shifted to the idea of chemical tags to identify the point of origin for smuggled drugs—a tactic even more closely tied to his absorbing work with the drug control program of the League of Nations. Not until the chance to apply substantial resources to the addiction problem came before a pharmacologist already dedicated to building a drug-development infrastructure in the United States did this research approach take on real significance. This occurred when Dunham approached the National Research Council in 1928 with a proposal that it take over coordination of scientific research on opiate addiction, as discussed in the next chapter.

With the committee's emphasis shifting ever more toward enforcement, Charles Terry's influence waned. As he continued work on the survey of opiate prescriptions in Detroit, the seventh of the city surveys, he made some attempts to be more tolerant of the views of committee members who disagreed with him.[48] It became clear, however, that Dunham and Terry disagreed too sharply to work together. An abortive plan to merge the committee with the American White Cross Anti-Narcotic Association, based in Seattle, would, if realized, have given Terry collaborators who favored addiction maintenance. Instead, the transfer of the scientific work to the National Research Council provided a means of shifting the bureau's resources in a direction Dunham approved of. When specific arrangements with the NRC were being worked out, Dunham asked that Terry be given some opportunity to remain with the committee rather than being summarily dismissed, but he undercut even this muted vote of confidence by adding that he was unable to assess the quality of Terry's work and could only vouch for the man's conscientiousness.[49] The disingenuousness of this remark is revealed in later correspondence with the NRC, when Dunham professed himself well qualified to endorse a research plan based on his long years of experience in the addiction field.[50]

As late as 1930, Terry continued to maintain that most addicts were not criminals, and in 1931, he again criticized federal drug control policy and spoke in favor of maintenance clinics.[51] After completing the Detroit survey, he retired from his position as executive to the Committee on Drug Addictions and never returned to addiction work.

Charles Terry represented the best of the humane approaches to opiate addicts in the 1910s and 1920s, and in some ways he was well placed to implement his views. That he failed to do so indicates how profoundly the tide ran against him as policy consolidated around prohibition of opiates, criminalization of addiction, and the institutionalization of addicts. No one could have successfully borne the standard of humane medical care for addicts in this period, as the composition of the addict population shifted toward types who would attract virtually no political support.[52] Even most medical discussion of addiction carried moral overtones, as in the distinction between more and less culpable addicts. Addiction treatment in the late 1920s consisted largely of managing the withdrawal process, with little follow-up, and any relapse into opiate use was considered added evidence of addiction's intractability rather than of the need for further research into treatment methods. As physicians in general practice became increasingly reluctant to treat addicts, the growing influence of a psychopathic model of addiction gave them a medical reason to refuse to accept junkies as patients.

3

The Technological Fix

The Search for a Nonaddicting Analgesic

As hopes of effective intervention dimmed, addicts became patients that physicians did not want to treat. Physicians had been blamed from within and without the profession for causing alarming levels of iatrogenic addiction; they were exhorted in therapeutics textbooks to avoid creating new addicts by reducing administration of opiates; and they were monitored by a federal bureaucracy ready to hand down indictments for excessive prescribing. From the late 1920s on, the American Medical Association cooperated closely with federal and state authorities to identify physicians addicted to opiates or who improperly prescribed them, and published the names of physicians convicted under the Harrison Narcotic Act in its *Journal*.[1] As gatekeepers to the only legitimate means of securing opiates, physicians became increasingly wary of patients complaining of mysterious pains and demanding relief through drugs.

As addicts became undesirable patients, the motivation to study the physiological mechanisms or human dimensions of addiction waned. Repeated treatment failures undermined hope that further study would yield a "cure" more meaningful than bringing the patient through the withdrawal process to an often fleeting state of abstinence. From the research at Philadelphia General Hospital, Light and Torrance's conclusions that chronic administration of opiates created little damage to organ systems and that withdrawal was not in itself a dangerous process (and thus did not require medication to improve patient safety) indicated no obvious directions for further study of managing withdrawal or investigating how opiates worked in the body. Lawrence Kolb's work (discussed in Chapter 5) branded chronically relapsing addicts as psychopaths and distinguished them from the normal "accidentals" who were addicted through medical treatment of long-term painful conditions such as cancer. The Treasury Department's Narcotic Division recognized this distinction and allowed maintenance for a few old or very sick addicts, while vigorously enforcing the law against unauthorized use of opiates. From this point of view, it was faulty character, not the drugs themselves, that seemed to be the seat of the problem, and support for lengthy incarceration as the right response culminated in the passage of the Porter Act to create the Public Health Service Narcotic Hospitals. Treatment institutions, possible sites for research on treatment methods, declined in number, and addicts increasingly faced criminal sanctions rather than medical compassion for their condition.

Finally, no front in physiological research seemed to offer promising leads for understanding addiction. This would change in the 1970s, when the discovery of opiate receptor sites and endogenous opioids and the use of brain-imaging technologies opened new avenues for studying the effects of opiates and other psychoactive drugs in the brain. In the 1930s, however, neurological research was concentrated on the transmission of impulses along the nerve cell, and whether nerves communicated via chemical or electrical signals was an ongoing debate into the 1940s. While receptor theory was guiding some research, this was focused primarily on hormonal action rather than on communication in the central nervous system.

The waning of research interest on these fronts helped create the context for substantial investment in an idea that had come to Lawrence Dunham's attention in 1926: that, just as novocain had reduced the medical need for cocaine, so a substitute for morphine that relieved severe pain but lacked addictive properties might provide a solution to the problem of opiate addiction. Dunham had quickly moved from this idea to the notion of tagging batches of morphine to indicate their manufacturing source, but his ongoing interest in scientific or technological solutions to what he saw as a problem arising from technological advance made him receptive to the idea of seeking a nonaddicting substitute for morphine when it was presented to him from another quarter.

A medication that would obviate the need for morphine promised to satisfy many interests. Physicians would welcome an analgesic they could administer without fear of addicting patients. If morphine could be eliminated from the pharmacopoeia, enforcement officials would be spared the cumbersome task of estimating the nation's legitimate medical need for opium and focus more simply on keeping all addictive opiates out of the country. The psychiatric explanation of addiction solidified by Lawrence Kolb argued that many addicted individuals might have lived marginally normal lives if not exposed to opiates. This idea reinforced the desirability of eliminating any availability of these drugs. In addition, the project of developing an improved analgesic created an opportunity that pharmacologists interested in building a pharmaceutical research and development infrastructure in the United States were able to exploit. From the outset the project to develop a nonaddicting analgesic was seen as consistent with the criminalization of opiate use and was structurally linked to the federal enforcement apparatus. These links, in turn, helped shape the scientific research and illustrate the intricate interactions between science and society.

Reid Hunt's Vision for American Pharmacology

"The field of medical research in which the United States is most conspicuously backward is that concerned with the discovery of drugs." With these words, Reid Hunt, professor of pharmacology at Harvard University, opened a letter to Victor C. Vaughan, chairman of the Division of Medical Sciences of the National Research Council (NRC), dated July 10, 1922. Yet, Hunt continued, pharmacology was a field of "vast importance" for public health, for the cure of disease, and for the relief of pain. Furthermore, he claimed, "There probably has never been a time when the outlook for the discovery of new drugs was so bright." However, nearly all the important drug discoveries of the preceding sixty years had been made in Germany. Although the recent world war had sparked American work in organic chemistry, Hunt noted, workers and facilities to test promising new therapeutic compounds were lacking in the United States. His purpose in writing was to solicit Vaughan's cooperation in urging the NRC to promote the development of American pharmacology. In a long list of specific suggestions, Hunt cited "the urgent need of nonhabit forming opiates and local anesthetics so that the use of opium and cocaine (the abuse of which almost balances the benefits) may be restricted or abolished."[2] In so doing, Hunt explicitly linked drug-development research to the public health problem of the nonmedical use of opiates—the same problem that the wide-ranging projects funded by the Bureau of Social Hygiene were intended to help solve.

Concerns about opiate addiction as a social problem created an opportunity to promote Hunt's program for building a drug-development infrastructure in the United States, as Bureau of Social Hygiene resources formerly distributed across diverse projects were shifted to a single enterprise: the search for a nonaddicting analgesic to replace morphine in medical use. At the same time, this single project brought a complex array of individuals, organizations, and interests into mutual cooperation in service of a set of interlocking aims. In 1928, the NRC appointed its Committee on Drug Addiction to seek a nonaddicting substitute for morphine. Through the research organized and overseen by the committee, and its links to a diverse set of scientific, professional, and policy interests, science and policy affected each other. In the process, drug prohibition contributed to shaping scientific research.

As Timothy Lenoir has argued, science is a cultural activity linked "in a seamless web" with social and political concerns and practices.[3] The Committee on Drug Addiction's structural links to the federal drug policy bureaucracy, to university research departments, to the American Medical Association, and to the pharmaceutical industry, as well as the influences these groups exerted on the scientific research overseen by the committee, illustrate just how science is embedded in and part of a larger social world. A network of overlapping groups agreed on drug development as a useful approach to solving the social problem of addiction, creating a mosaic of interests within which each group achieved specific internal gains while offering support to a larger effort.[4] The parties involved included academic chemists and pharmacologists, philanthropies and social reformers, organized medicine as represented by the AMA, pharmaceutical firms, clinicians, public health officials, and law enforcement officials. A research problem that was compelling to all of these constituencies had the ingredients to secure durable funding and maintain long-term research interest. The quest for a nonaddicting analgesic thus functioned as a "boundary object" in the sense portrayed by Susan Leigh Star and James Griesemer.[5] The project represented a meaningful objective for all the participating groups and thus provided a focus for coalition and collaboration. At the same time, it embodied different kinds of interests and different internal meanings within each group, so that each could pursue its distinct aims, while remaining allied in pursuit of the shared objective.

The NRC Committee on Drug Addiction

Victor Vaughan responded favorably to Hunt's 1922 proposal that the NRC exert leadership in developing American pharmacology; he urged Hunt to submit a research plan and budget that Vaughan could present to the executive committee of the NRC's Division of Medical Sciences in the fall.[6] Hunt proposed the creation of a pharmacological research institute analogous to the Carnegie Institution's Laboratory for Embryology at Johns Hopkins University; he envisaged research pharmacologists, free from burdensome teaching duties, housed not in a medical school (where, he said, too few faculty members were sufficiently schooled in chemistry or physics) but in a graduate school. There, pharmacologists would be able to consult freely with chemists who developed compounds, physiologists and biochemists who studied chemical processes in the body, immunologists, and so forth.[7] At its meeting of September 22, 1922, the executive committee of the Medical Sciences Division moved that Hunt prepare a more detailed report to be used as the basis for seeking $25,000 to $50,000 a year to fund pharmacological research.[8] At this point, action stalled. Vaughan had been replaced as chair of the executive committee, and Hunt left the country for most of 1923 to teach at the Peking Union Medical College. A Committee on Pharmacological Research was created; it sent questionnaires to every pharmacology laboratory in the United States to solicit views on how the NRC might profitably support pharmacology. Results were compiled in 1928, just as an offer of substantial funds presented the NRC with the impetus to select a particular pharmacological focus.

In 1928, Lawrence Dunham replaced Katharine Bement Davis as head of the Bureau of Social Hygiene. With his background in law enforcement, Dunham's chief interest was in participating in the treaty-making process headquartered at Geneva and in shifting the bureau's work toward criminology. Under his leadership, it was decided to transfer the scientific work of the Committee on Drug Addictions—consisting of research at various sites into the physiological effects of opiates and treatment of addiction, as well as assessment of amounts of morphine necessary to meet true medical needs—to other suitable auspices.

In an overture to the NRC's permanent secretary, Vernon Kellogg,[9] on November 23, 1928, Dunham described the addiction research funded by the Bureau of Social Hygiene as "a purely scientific undertaking" that included "most of the leading men in the country in this field," and that represented "the most fundamental work that is being done anywhere in this field." He characterized the committee's wide-ranging objectives as follows: "The purpose is to find out what morphinism is, what the drug does to the human body to cause addiction, and what if anything can be done by way of prevention or cure."[10] The only thing lacking, Dunham said, was "administrative leadership and control by some permanent organization such as your own." Would the NRC consider taking over this work and providing the required leadership?

Kellogg passed this letter to William Charles White, a pathologist in charge of tuberculosis research at the Public Health Service's Hygienic Laboratory, forerunner of the National Institutes of Health. White was familiar with work sponsored by the Bureau of Social Hygiene, and he did not share Dunham's enthusiastic appraisal: "It has not been very fruitful in advancing our knowledge in this field," he wrote Dunham. White was more impressed with Lawrence Kolb's theory that, as White put it, "morphinism is coupled always with some deficiencies in the person who becomes addicted to it."[11]

Like Kolb, White appeared to believe that the presence of these predisposing psychological factors obviated the need for further studies of the etiology of addiction. White expressed cautious interest in Dunham's offer and then polled the members of the Division of Medical Sciences as to whether the NRC should accept the proffered funds. Approvals from a majority plus a green light from the NRC's executive board soon followed. In December 1928, the NRC appointed a committee—to be called the Committee on Drug Addiction—to administer the project.[12] Its original members, under White's chairmanship, were Reid Hunt; his fellow pharmacologist Carl Voegtlin of the Hygienic Laboratory; Frederick B. LaForge, a chemist in the insecticide division of the U.S. Bureau of Chemistry and Soils; and C. S. Hudson, a Hygienic Laboratory chemist.[13] Among those who soon joined the committee and remained members throughout the 1930s was the eminent pharmacologist Torald Sollmann of Western Reserve University in Cleveland.[14]

Realizing the Vision

Hunt seized the opportunity to advance the research agenda he had laid out in 1922. In a letter to Voegtlin, he noted that the introduction of such drugs as the salicylates and colchicum had already reduced the number of medical conditions for which morphine was the only or the best available drug. He then laid out a plan of action based on the search for new drugs: "The development of local anesthetics might serve as a precedent for further work on opium. A careful study of the cocain [*sic*] molecule showed that only certain [structural] groups are essential for local anesthesia; nearly all of the other groups were whittled away until now we have so many local anesthetics that there is little excuse for a physician to use . . . cocaine." Hunt suggested that similar study and modification of the morphine molecule might yield new compounds that would similarly pare away undesirable effects like addictiveness while retaining desirable ones like analgesia. Desired effects might further be separated from each other, so that individual drugs could be targeted to individual body conditions. Such a project, in Hunt's view, would require "cooperation between the highest type of organic chemists and pharmacologists" and would probably interest pharmaceutical manufacturers as well. Drug-development work along these lines would be more useful than "additional studies on the mechanism of the morphine habit, efforts to improve the treatment of addiction, etc.," Hunt concluded—in other words, than the kinds of work the Bureau of Social Hygiene had funded.[15] Although the bureau's committee had also included eminent pharmacologists and chemists (such as Alfred Newton Richards of the University of Pennsylvania and Lafayette Mendel of Yale), and it had funded some laboratory research (such as Oscar Plant's at Iowa), its funding approach had been scattershot. Hunt, in contrast, proposed a focused, coordinated quest for a nonaddicting opiate analgesic.

Hunt persuaded his fellow committee members to devote the bureau's grant to this search, in part by presenting it as a promising approach to the addiction problem. Hunt's prominence in various settings related to drugs and medicine undoubtedly made him realize that funding the development of a nonaddicting analgesic promised to satisfy a variety of interests. As a member of the AMA's Council on Pharmacy and Chemistry and the U.S. Pharmacopoeia Commission, he was aware that physicians, as part of a larger campaign to rid the marketplace of fraudulent drugs and gain control over the marketing of medicines, sought to distance themselves from opiates, as old-fashioned medicine, and thus from iatrogenic addiction. And his membership on the Treasury Department's Special Narcotic Committee in 1919 acquainted him with the alarmingly high (and exaggerated) estimates of numbers of opiate addicts that were fueling federal concern about addiction. Hunt also argued the value of such a project in advancing alkaloidal chemistry and pharmacology in the United States. In doing so, he yoked the Bureau of Social Hygiene committee's mission to the aspirations of American pharmacologists.

American Pharmacology in the 1920s

In the decade following World War I, pharmacologists in the United States faced challenges and opportunities arising both from the war itself and from the ongoing reform of American medical education. When hostilities had cut off access to German products, American dependence on German pharmaceuticals had become starkly apparent. Hunt's 1922 appeal to the NRC had not been an isolated effort to promote drug development in the United States. In the early 1920s, a group of American chemists and pharmacologists, led by the chemist Charles Holmes Herty, and including Hunt, formed a Committee on an Institute for Chemo-Medical Research. Although this effort failed to establish the proposed institute, as funds for an endowment could not be found, the committee's widely disseminated report promulgated a vision of cooperative research in the quest for improved medicines, much as Hunt had done in his letter to Vaughan.[16] The cooperative aspect was based on two institutional models, one German and one American.

In Germany, close links between chemists and pharmacologists, and between academic and industrial scientists, had led to the dramatic discoveries Hunt envied. Paul Ehrlich's development of Salvarsan, a syphilis treatment significantly more effective than any previously known, exemplified the vision of the "magic bullet," a powerful medicine that attacked the invading organism causing infectious disease. The phrase "magic bullet" rhetorically expressed the aim of a research program based on the assumption that a drug's effects were a direct function of its molecular structure.[17] Thus, as organic chemists in Ehrlich's Institute for Experimental Therapy in Frankfurt produced novel compounds, pharmacologists tested them in animals to determine their range of physiological effects. Among hundreds of test compounds, one might emerge that, like Salvarsan, was a significant improvement over existing medications. Compounds that failed to show therapeutic benefit nevertheless yielded useful information. Chemists and pharmacologists correlated molecular structures with their physiological effects to produce data on relationships between structure and activity.[18]

Despite hopes of a predictive theoretical basis, drug development proceeded by largely empirical means to utilitarian ends: the discovery and production of new medicines. Results occasionally verged on the miraculous. Hunt cited several examples, including quinidine, which could, "within a few hours, convert an absolutely irregular heart into a perfectly normally beating one."[19] More often, however, new drugs were deemed worth developing because they demonstrated some moderate improvement over medications already in use. Analysis that weighed therapeutic benefits against the drawbacks of toxicity was essentially also a market analysis: even limited comparative advantage over existing remedies could warrant the substantial investment in industrial production of a substance hitherto made only in small batches for research purposes. In this way, the financial interests of pharmaceutical companies moved in tandem with the scientific interests of academic pharmacologists. For practicing physicians, too, marginally better medications were welcomed as improvements over existing remedies. All three groups, then, foresaw benefits from research into drug development.

Americans who spent time at Ehrlich's laboratory, as Reid Hunt did in 1902 and 1903, were inspired to pursue such research when they returned home.[20] The U.S. Army Chemical Warfare Service provided an American precedent for tightly organized cooperative research to identify useful compounds and produce them in large quantities. Although the objective was to find toxic agents for use in combat, the methods resembled those by which German scientists and industrialists launched new medications. Compounds produced by chemists were immediately tested by pharmacologists, and the requirements of large-scale production were investigated as soon as one seemed promising. Hunt was one of the thousands of scientists involved in the work of the Chemical Warfare Service, and after the war, he became one of a small number of influential chemists and pharmacologists who vigorously promoted cooperative, project-centered research to develop new medicines.[21]

Pharmacology in the 1920s was one of several biomedical disciplines given stimulus by the reform of American medical education. The move to place medical education on a scientific foundation prompted the transformation of traditional teaching in materia medica (lists of remedies with their preparation methods and their indications) to scientific pharmacology. As medical school pharmacology departments were created or transformed, service teaching roles emerged, but corresponding support for research was lacking. In this respect, pharmacology resembled the related disciplines of physiology and biochemistry and faced similar challenges in developing a strong institutional base for an independent discipline.[22] However, tensions peculiar to pharmacology lay in its relation to therapeutics in clinical practice and to the manufacture of medications.

The American Pharmaceutical Industry

In contrast to their German counterparts, few American pharmaceutical firms before World War I extended their scientific efforts beyond the quality-control tasks of maintaining purity and dose standardization. The industry as a whole, moreover, was tainted by long historical association with unscrupulous manufacturers of nostrums. Forging an alliance with academic pharmacology would provide a means of gaining useful compounds, pertinent research findings, and the legitimacy conferred by association with university-based scientists. Leaders of ethical pharmaceutical firms such as Eli Lilly and Company believed that the future lay in research-based development of new drugs targeted to specific illnesses; firms like Merck and Company, formed when ties were cut between American branches and their home companies in Germany, sought to transplant the German research tradition to American soil. In the 1920s, these and other leading pharmaceutical firms began to form research departments and to forge cooperative research links with academic scientists.[23]

Despite this growing cooperation, the suspect status of the pharmaceutical industry was reflected in a lingering fear that association with commercial drug manufacturers would discredit the scientific pretensions of pharmacologists. Unlike the national associations for related disciplines such as physiology, the American Society for Pharmacology and Experimental Therapeutics (founded in 1908 by John J. Abel) refused membership to employees of commercial concerns, regardless of their education qualifications or research accomplishments. The membership ban remained in force until 1941, despite a widespread feeling among university-based pharmacologists by the late 1920s that some of their colleagues in industry were high-caliber scientists who deserved membership.[24]

Disciplinary Aspirations

Physicians had traditionally considered materia medica an empirical adjunct to therapeutics.[25] To claim the status of a scientific discipline, pharmacology needed to generate new and important knowledge through laboratory studies. John J. Abel, widely considered the founder of American pharmacology, expressed the discipline's scientific aspirations in 1891: "Briefly this science tries to discover all the chemical and physical changes that go on in a living thing that has absorbed a substance capable of producing such changes, and it also attempts to discover the fate of the substance incorporated. It is not therefore an applied science, like therapeutics, but it is one of the biological sciences, using that word in its widest sense."[26]

Yet, as Hunt had noted, the promise of developing useful and perhaps revolutionary new treatments also represented a potential source of legitimacy. These divergent visions reflected two research strains within pharmacology. A physiological strain, growing out of the work of the French scientists François Magendie and Claude Bernard, emphasized drug administration as a means of studying physiological processes in living organisms—to elucidate normal functioning by observing its response to different drugs, or to determine the mechanisms of action of individual drugs. An organic chemistry strain aimed at medication development by such methods as those pursued by Ehrlich.

In laying out his program, Hunt balanced both views of pharmacology. On the one hand, he saw organic chemistry as the "basis of modern pharmacology." On the other hand, acknowledging that most of the recently discovered medications had emerged from work that was "for the most part haphazard," he called analysis of mechanisms of action "real pharmacological studies." He linked these ideas by implying that further study of mechanisms of action would enable pharmacologists to predict with greater probability of success which compounds would prove useful. That Salvarsan had been the 606th in a series of compounds Ehrlich developed and tested was well known among both scientists and physicians.[27] Hunt said that enough had since been learned about its mechanism of action "as to suggest that probably 90 percent of the work done on related compounds had from the beginning almost no chance of success."[28] He thus implied that such studies would enable future drug-development efforts to be effectively guided by theory; in the meantime, the state of knowledge was such that somewhat empirical means were still necessary. Thus, studies of physiological functioning, as affected by drugs, held the scientific high ground. Yet, the results of drug-development studies, despite their association with pragmatic ends, also formed respectable scientific findings.

For such studies, understanding the mechanism of action was not necessary; rather, pharmacologists administered test compounds in precisely measured doses to animals and observed specific physiological responses, such as changes in blood pressure or respiratory rate. Through such bioassays, they determined toxic and therapeutic effects, noted dose ranges in which these appeared, and correlated such effects to the molecular structures of the test compounds. In the search for new medicines, evoking a desired therapeutic response without incurring unacceptable toxicity was more important than understanding the mechanism of that response. These methods constituted the core of what pharmacologists today refer to as "classical pharmacology," as drug-development research diverged from physiological studies. Dose-specific profiles of individual drug actions, in turn, reformulated therapeutics in the new language of laboratory science.

Drugs and Therapeutics

From the point of view of medical practice, pharmacological research in this period can be seen as one facet of a transformation of drugs in medicine, as orthodox physicians wrested control of medical knowledge and practice from contending medical sects and from lay healers (including patients themselves). Traditionally, medications had been intended to affect the entire system by readjusting body fluids. Cathartics and emetics, with their visible discharges, were emblematic of this approach.[29] Meanwhile, in the unregulated medication marketplace of the nineteenth century, people often medicated themselves. Curative benefits were not clearly differentiated from symptomatic relief or simply feeling better. The separation between "medical" and "recreational" uses of drugs, a familiar feature of current discourse, had not yet been constructed.[30] This flux in the meanings of drugs and medicines also shaped the context in which many individuals in this period began their use of morphine and heroin and started on the path to addiction.

Such a distinction emerged in part as activist physicians gained control over access to medicines and over the right to determine the limits of "medical" uses of drugs. From the 1870s, the AMA had actively worked to disprove the extravagant claims of the purveyors of nostrums, submitting their products to chemical analysis and exposing false claims as to ingredients and effects. Following a reorganization in 1901, the AMA emerged as a powerful public relations and lobbying force in the interests of private practice physicians.[31] In 1905, it created a Council on Pharmacy and Chemistry to analyze new drugs and make recommendations regarding their appropriateness for medical practice.[32] Lists of those it deemed useful were published each year in the AMA's *New and Nonofficial Remedies,* which kept physicians informed about new medicines pending their acceptance in the *United States Pharmacopoeia,* a slower process. These activities gave physicians increased power to determine what substances could be considered medicines and how and by whom they could be used. They also implied recognition that a science-based pharmacology would produce an ongoing stream of new medications to be assessed for appropriateness in medical practice.

American physicians had reason to desire improved opiates in particular, given the widespread view, within and without the profession, that physicians' overprescription of opiates was chiefly responsible for a disturbing prevalence of addiction among the general population.[33] By the late 1920s, physicians had substantially reduced medical use of opiates, and the demographics of opiate addiction had shifted away from patients. Nevertheless, the AMA worked actively throughout this period to free physicians from the taint of causing addiction. It supported major pieces of Progressive Era drug legislation, including the 1906 Pure Food and Drugs Act and the 1914 Harrison Act. Although it challenged certain technical points in the Harrison Act, the AMA, at a time when it was jealously guarding the physician's prerogative to be the sole authority in matters of medical practice, made no protest against this incursion into the arena of professional judgement. It implicitly agreed with the Supreme Court's anti-maintenance decision when, in 1919 and 1924, it passed resolutions against the ambulatory treatment of addicts. Educating physicians to reduce medical uses of opiates by observing narrower indications for their prescription, reducing doses, and replacing them with newer, more specific drugs was part of the AMA's campaign to distance physicians from association with improper use of opiates more generally and from addicts specifically.

The Committee Begins Its Work

Given these wide-ranging reasons for supporting drug development in general and for seeking a nonaddicting opiate analgesic in particular, it is not surprising that Hunt persuaded the committee to decide unanimously to devote the Bureau of Social Hygiene grant to the search for a nonaddicting analgesic opiate. In the style of Progressive Era reformers, Hunt presented this as a way that scientific research might produce the solution to a social problem, but given his long-standing advocacy of pharmacological research, his statement that the project would advance alkaloidal chemistry and pharmacology in the United States undoubtedly meant more to him. The latter vision, calling for coordination across disciplinary lines and for investment in an infrastructure to support basic research, was in tune with other developments in American science in the 1920s, a period when private philanthropies such as the Rockefeller Foundation and the Carnegie Corporation were creating new systems of science patronage and organization.[34]

Moreover, Hunt's proposal meant pulling funds from a variety of BSH-supported projects, some of which had addressed treatment and prevalence issues, and funneling all the resources into laboratory-based research. This move directly paralleled the move to NRC auspices of another component of the Bureau of Social Hygiene's work, with a similar shift away from social concerns toward a focus on laboratory research. The

BSH had originally formed to study prostitution and related urban problems, and, before 1920, it had funded both research and projects intended to have direct social impact; for example, the BSH provided grants to the American Social Hygiene Association, an organization seeking to expand and improve sex education as a means of reducing incidence of venereal diseases. In 1921, the BSH proposed to the NRC that it create and manage a committee to supervise relevant scientific research. Just as Hunt would later sense the opportunity to capture proffered funds for the development of pharmacology, the biologist Frank Lillie led a successful effort to devote BSH resources to laboratory-based endocrinological research on reproduction under the auspices of the NRC's Committee for Research on Problems of Sex. Funds were thus directed away from the study of human sexuality and its social context and used to further the disciplinary aims of reproductive endocrinology. As Adele Clarke has shown for reproductive science, and as I argue with respect to opiate pharmacology, in both cases, the move to the NRC created a context in which scientists could capture problems associated with social stigma and transform them into research problems that constituted basic research, while also furthering disciplinary aspirations.[35] In the case of opiate pharmacology, the focus on developing improved analgesics proved remarkably compatible with the stiffening orthodoxy of federal policy in the classic era of drug control.[36]

After preliminary discussions with individuals at several universities, the newly formed Committee on Drug Addiction chose sites and engaged scientists to begin the work. The components would include a chemistry laboratory to generate test compounds, a pharmacology laboratory to conduct animal tests, and clinical sites for human studies. The University of Virginia was chosen as the site for the chemistry laboratory because of the presence there of Lyndon F. Small, a Harvard-trained organic chemist who had spent two years in the laboratory of Heinrich Wieland in Munich working on the chemistry of opium alkaloids (first as a Sheldon Traveling Fellow from Harvard and then as an NRC fellow).[37] The committee recognized Small as virtually the only American familiar with the chemistry of these compounds. He immediately began seeking a European chemist to bring to the laboratory, emphasizing the lack of American experience in this area. After polling his European contacts as to who might be available, he offered the position to Erich Mosettig of Vienna. Small developed a two-pronged research plan: he would work on the morphine molecule, breaking it down by hydrogenation and adding new groups at various points on the molecular structure to develop new morphine derivatives. Mosettig would take a synthetic approach: starting with phenanthrene, a coal tar product that matched the structural core of the morphine molecule, he would build new synthetic structures.

Because the goal was a body of knowledge correlating molecular structure to observed pharmacological activity, Small intended systematically to cover a logical series of modifications of the morphine structure. This strategy meant examining some compounds that had already been developed elsewhere. His first task, assigned by the committee, was to survey the chemical literature on opiates; here, he uncovered discrepancies and errors regarding molecular structure of numerous compounds. By proceeding systematically, characterizing each compound in terms of its physical properties and making cogent arguments for the precise structure of each, Small hoped to bring greater order to the field. This strategy also implicitly staked a claim to an area of work that was of interest to other laboratories. As the work progressed, Small corresponded with other chemists in the United States, Germany, England, and France, informing them of his laboratory's intentions to work and publish in various specific areas, and at times working out agreements regarding credit or priority of publication.

The committee originally arranged with Treat Johnson's chemistry laboratory at Yale to work on the morphine molecule; Johnson put Richard Manske to work on this project. Small was clearly unhappy with this division of the work; he wanted to be closely in touch with every aspect. He further complained that Manske was not forthcoming about the progress of his work. The Yale grant was not renewed, and the committee cited the need to concentrate the work at a single location where absolute uniformity of methods could be maintained to assure the purity of each substance produced. The importance of such uniformity became increasingly clear when difficulties arose in making larger quantities of promising substances; this factor helped keep all the planned work at a single set of coordinated institutional sites. This was one of many factors helping maintain the boundaries of the project, keeping it in the purview of the NRC committee's investigators and discouraging others from working in the same area. Others included the committee's decision not to allow the project chemists to distribute compounds to laboratories outside the committee's own, except when no possible overlap with the committee's work would occur.

White arranged for the pharmacological investigations to be conducted at the University of Michigan under the supervision of Charles W. Edmunds. Edmunds hired Nathan B. Eddy to set up the laboratory and test the various compounds for therapeutic effect and for toxicity, including the potential to produce addiction. For compounds deemed promising enough for human studies, the Public Health Service would make available its facilities at the federal Prison Annex at Fort Leavenworth, Kansas, and, after its opening in 1935, at the Lexington Narcotic Hospital in Kentucky. The NRC's earlier Committee on Pharmacological Research was absorbed into the Committee on Drug Addiction in November, 1929.[38] In 1933, when Alfred Newton Richards shifted his research in another direction, the BHS-funded work at the University of Pennsylvania was moved to Edmunds's laboratory.[39]

Federal Involvement

From its earliest stages, the committee sought and obtained cooperation from federal officials active in drug enforcement and addiction research. These links kept the work tied to policy that reflected the new, harsher view of addicts; they also suggest that the quest for the nonaddicting analgesic was meaningful to these groups as well. From the outset, White sought to link his committee's work to the planned Public Health Service narcotics hospitals called for in the Porter legislation, or, as he put it, "the new Government plan for the segregation and care of Federal addicts."[40] Representatives of two federal agencies were added to the committee: Commissioner Harry Anslinger of the Bureau of Narcotics and Walter Treadway, chief of the Public Health Service's Division of Mental Hygiene. This division was created specifically to build and manage the two Public Health Service narcotics hospitals authorized by the Porter Act of 1929.[41]

Through Anslinger, large amounts of confiscated morphine were made available to Small's group; these stores provided the starting material for many of the compounds they synthesized, both initially as novel structures and for further testing in larger quantities as needed. In addition, Anslinger's presence on the committee, like the connections with the Bureau of Social Hygiene, linked the work to the international supply-side drug control policy structure. The goal of developing a nonaddicting substitute for morphine was entirely consistent with the policy aim of reducing the incidence of addiction by restricting manufacture and distribution of opiates to medically authorized channels. As the work progressed, Small became an occasional expert consultant to the American delegation to Geneva and also to the League's Opium Advisory Committee. Treadway, the Public Health Service psychiatrist in charge of establishing the prison hospitals that would connect the medical and criminal justice responses to addiction, was to oversee all clinical testing of compounds that emerged from the animal tests with sufficient promise. Locating the clinical testing efforts at these institutions would mean that otherwise healthy men who were addicted to morphine or heroin would be available for studies of new compounds. As will be seen, these subjects proved to be crucial resources as soon as the question of a test drug's addictiveness arose. The seamless blending of the laboratory research effort and federal policy aims is seen in this retrospective statement of the purpose of the committee's research efforts that appeared in a 1937 report on committee activities: "The main purpose would be to replace, if possible, as novocain replaced cocain, all the uses of morphine with better or equal drugs not having the addiction property. That would then force the addiction problem into the police field of which Commissioner Anslinger has charge."[42]

Surveying the Pharmacological Literature

Among the early tasks set by the committee was a survey of the pharmacological literature on opiates comparable to Small's survey of the chemical literature. Before arrangements to locate the pharmacological work at Michigan were completed, the committee selected the Cornell pharmacologist Robert Hatcher, a former student of Torald Sollmann, for this task. An original member of AMA's Council on Pharmacy and Chemistry, Hatcher was a prestigious clinical researcher and advocate for more systematic study of new medications.[43] He brought the laboratory pharmacologist's perspective to bear on the pharmacological literature on opiates, whose clinical aspects reflected attempts to assess drugs by administering them in clinical situations rather than in controlled laboratory studies. He recognized that dose, duration of action, and route of administration all affected the range and intensity of observed drug effects. Just as Small's survey of the chemical literature was uncovering errors in the structural models proposed for various opiate molecules, Hatcher's survey revealed a field in disarray, filled with conflicting or inconclusive reports on many points, especially on whether particular drugs (such as codeine, an opium derivative, and Dilaudid, a recently introduced morphine derivative) were addicting. Again, the committee saw in this confusion in the literature further justification of its programmatic aims. Hatcher concluded that impurities, and inconsistencies in dose amounts or dose intervals, were likely explanations of some of the confusion. The arguments for concentrating the work at a single chemistry laboratory, where consistency of product could be maintained, and a single pharmacological laboratory, where consistency of method and measurement could be assured, were reinforced.[44]

Hatcher also urged the committee to adopt a goal of determining the minimal indispensable uses of opiates in medical practice. Such a project would link the committee's work to ongoing concerns of the AMA. By 1930, the presence of a substantial federal enforcement operation against opiates, the growing influence of psychiatric theories that personality defects caused chronic addiction, and the plans to construct Public Health Service prison hospitals for addicts all reflected a growing perception that opiate addiction was associated with urban sectors where vice was common. Nevertheless, the AMA continued to be motivated by the stigma physicians had suffered as overprescribers of opiates. Hatcher, working in concert with Morris Fishbein, then editor of the *Journal of the American Medical Association,* developed a series of articles on indispensable uses of opiates. These were published in *JAMA* with accompanying editorial messages; the AMA later brought them out as a book.[45]

By late 1930, both the chemical and pharmacological laboratories were established and at work. At Michigan, Nathan Eddy and his team ran each new compound supplied by Small's group through a broad array of assays for analgesic, respiratory, depressant, hypnotic, intestinal and other effects. Eddy worked on analgesia—developing, for example, devices and methods for quantifying pain levels. Neither clinical tests nor tests to determine addictiveness of the test compounds were yet under way. These were to be devised as the promise of some test compounds was deemed to warrant them.

Desomorphine: Dashed Hopes, Methodological Advances

One of the early promising compounds developed in Small's laboratory was dihydrodesoxymorphine-D, later named Desomorphine. As this drug moved through pharmacological and clinical trials, early hopes that it represented a nonaddicting analgesic substitute for morphine were dashed by later confirmation that it was highly addictive. Following this test drug through these studies illustrates how the committee and its researchers worked together and points up the methodological problems confronting researchers trying to assess addictiveness.

Small directed the synthesis of the dihydrodesoxymorphine-D molecule in January 1930; once he perfected a method for producing sufficient (although still minute) quantities, he sent samples to the pharmacological laboratory at Michigan, where Eddy undertook the animal studies.[46] Initial results showed that Desomorphine had significant analgesic power, and committee members expressed hope that it was the drug they were seeking.[47] They recommended that clinical tests be undertaken under the supervision of Walter Treadway of the Public Health Service.

Before the planned clinical tests could begin, a technical hurdle had to be surmounted. The clinical tests would require larger amounts of the compound than Small could make in his own laboratory, and so he turned to the pharmaceutical industry to manufacture the necessary quantities with the requisite uniformity and purity. Although Small was getting ample supplies of morphine from the Bureau of Narcotics, and morphine could easily be converted to codeine as needed, the research plan also called for studying and modifying a number of other alkaloids occurring in raw opium. These he sought to obtain from manufacturers. Under the Harrison Act, only three U.S. pharmaceutical firms were permitted to import and process opium (although other companies could then buy opiates from these manufacturers for inclusion in their own products). These were Merck and Company, Mallinckrodt Chemical Works, and the New York Quinine Company.

Small's initial contacts with two of these companies, Merck and Mallinckrodt, blossomed into long-standing relationships of close cooperation. Both companies funded fellowships in Small's laboratory (in addition, E. R. Squibb & Sons funded one each in Small's and Eddy's laboratories). A steady stream of graduates moved from the University of Virginia program to positions in these companies, and Small was often involved in individual hiring decisions. He might alert a company to someone seeking the salary and security offered by a permanent job in industry in preference to a lower academic stipend that could be guaranteed only for the life of the current grant; or he might ask that a company defer an offer to a particular individual because he felt he couldn't do without the person's skills just yet.[48] In addition, Small and the committee's chairman, William Charles White, kept their close friends in the industry (most notably, George Merck, president of Merck and Company) apprised of laboratory developments and of confidential committee discussions.[49]

Immediately following the conference at which Eddy announced to the committee members and project workers the extraordinary analgesic potency of dihydrodesoxymorphine-D, Small wrote to Dr. Joseph Rosin at Merck informing him of the drug's promise: it was ten times stronger in analgesic power than morphine but had low toxicity. Based on what was known of the drug so far, Small compared it favorably with heroin and the recently introduced Dilaudid. Of the new drug's addiction properties, Small said, nothing was yet known.[50]

Other companies were quick to approach Small following any hint that a potential new therapy was being discovered. As soon as Eddy's findings on the analgesic strength of dihydrodesoxymorphine-D appeared in the academic press, several companies (including ones that had no previous connection with the project) requested permission to develop the drug for the market. Two factors excluded companies not already involved in the work from consideration: Small indicated that, naturally, those companies that had already made an investment in the project should have the first opportunity to develop any resulting products for market;[51] furthermore, because the companies that were already involved had employees, trained by Small and Mosettig, who were intimately familiar with the necessary alkaloid chemistry, they were in the best position to undertake scaling up production, necessary even for the limited quantities required for further research.

The legal status of the test drug also played a role. As a morphine derivative, dihydrodesoxymorphine-D fell under the Harrison Act. Committee member Harry Anslinger, commissioner of the Bureau of Narcotics, involved himself in all questions relating to control of availability, including manufacture of supplies for clinical testing. He initially envisaged an equitable arrangement in which each of the companies authorized to process imported opium would undertake the small-scale manufacture of test supplies of dihydrodesoxymorphine-D; however, the engineering difficulties involved in developing the process meant that such an arrangement would entail costly duplication of effort. Instead, he selected Merck to develop the process for making dihydrodesoxymorphine-D, with the understanding that the other companies would, in sequence, have the opportunity to work with other promising drugs to emerge from the research. Clearly, having the sole opportunity to devise a manufacturing process during testing would confer a substantial advantage should a marketable product result. Various aspects of drug development, such as creating new compounds, conducting animal tests, and even coordinating clinical tests, were carried out within both industrial and academic settings; however, the engineering task of developing processes for large-scale manufacture lay exclusively within the proprietary realm of individual manufacturing firms. The potential difficulty of devising such processes was emphasized when Frank Cohen, one of Small's hand-trained chemists, who had joined Merck, was unable to obtain sufficient yields of dihydrodesoxymorphine-D even with Small's guidance from Virginia. When written communications failed to resolve the technical problems, Small made several visits to the Merck plant in New Jersey in December 1934.[52] These difficulties delayed clinical testing until early 1935.

In his reply to Small's communication about the potential of dihydrodesoxymorphine-D, Joseph Rosin renewed an offer from George Merck to loan the company's patent attorneys to Small so that he could patent the new drug.[53] The committee agreed to let Small file an application but made no commitment to following through and securing the patent. During committee discussions, Anslinger suggested that patents might function as tools to enhance government control of addictive drugs. He asked whether the government could take out a secret patent on dihydrodesoxymorphine-D so that others could be prevented from knowing how to make it should it prove dangerous; he was informed that a patent, by its nature, is not secret. Lacking a means of controlling access to knowledge of how to make the drug, federal enforcement officials would have to focus, should the need arise, on trying to control access to the drug itself.

In 1924, the NRC had adopted a policy that called for it to secure patents on discoveries made under its auspices and to dedicate these patents to public use. With respect to any patentable drugs developed through work sponsored by the Committee on Drug Addiction, Stanhope Bayne-Jones, who was chair of the NRC's Division of Medical Sciences in 1933, acknowledged that the whole issue of patenting discoveries made in academic settings was problematic. Nevertheless, while granting that pharmaceutical companies that had materially aided the work had a legitimate interest in any resulting products, he asserted that this interest was superseded by that of the NRC.[54] Charles Edmunds, at Michigan, argued that any product royalties resulting from the project's researches should directly benefit the further work of the Committee on Drug Addiction. He cited as precedent an arrangement between the University of Michigan and Parke, Davis & Company, in which the latter took out a patent on a project developed with help from Michigan scientists and then paid fixed royalties on the product to support further research at Michigan.

Meanwhile, the NRC's Division of Medical Sciences (home of the Committee on Drug Addiction) formulated a policy that no useful product of researches it sponsored should be patented, since medications should be available without restriction to the sick people who needed them. This policy was at odds with that developed earlier by the NRC, which overturned it (the differences were tactical; both policies sought to devote useful products of research to the public good). These discussions reflect persistent tensions: physicians wanted to downplay association with the commercial aspects of the drug market while increasing control over drug access, scientists wanted to publish their results in the academic literature, and pharmaceutical firms wanted to retain proprietary control of information and its uses.[55] Patents allowed a means of resolving some of these tensions. By affording control of rights to license manufacture, they could protect proprietary interests but at the same time permit accounts of how to make and use new drugs to be published in scientific and medical literature.

A potentially addicting opiate remained a unique case, and in the end, thanks to the influence of Anslinger, the committee decided to have Small patent the process for making the drug and assign all rights to the secretary of the Treasury. This, the committee decided, would enable the government to retain control and act for the public good, since the Treasury could distribute licenses in such a way as to ensure adequate distribution at a fair price—or it could withhold license to manufacture altogether.

Concurrently with patent discussions and with the initial production efforts at Merck, the committee conferred with the AMA on selection of a trade name. The AMA's involvement in the development of new medications included, besides the activities already described, a Committee on the Pharmacopoeia, which advised on the inclusion of drugs in the *United States Pharmacopoeia,* and a Committee on Nomenclature, which coordinated selection of trade names for new medications. The Committee on Drug Addiction submitted dihydrodesoxymorphine-D to the former for approval and to the latter for registry under a trade name. The Committee on Nomenclature responded with the suggested trade name "Dysomorphine," based on elements of the chemical name. Committee members and scientists offered variants; finally, the committee adopted a Merck employee's suggestion, Desomorphine.[56]

While manufacturing snags and patenting details were being worked out, Walter Treadway was planning and initiating clinical studies. He envisaged two kinds of tests: analgesia and anti-cough tests with cancer and tuberculosis patients, and tests for addiction potential. He arranged with the Department of Health of the state of Massachusetts to have the analgesia and cough relief tests carried out at several state hospitals in Massachusetts. These studies quickly confirmed Desomorphine's superior analgesic strength as compared to morphine. Tests for addiction potential were to be conducted with addicted federal prisoners at the Fort Leavenworth Prison Annex. Here the Public Health Service had set up a preliminary research site, pending construction and opening of the planned Public Health Service Narcotic Hospital at Lexington, Kentucky. Treadway assigned Clifton Himmelsbach, a PHS physician, to oversee tests with addicted prisoners at Leavenworth and Lexington.[57] The committee sent Himmelsbach on a training tour that included two weeks each at the laboratories of Eddy and Small and six months with Torald Sollmann at Western Reserve University in Cleveland.[58]

As Hatcher's literature survey had indicated, the question of addictiveness was surrounded by confusion. There was no known reliable method for determining whether a drug was addictive before launching it into medical practice. Physicians and researchers recognized the syndrome that typically followed abrupt cessation of morphine use by an addicted individual. They also associated tolerance, the need over time to escalate doses to achieve the same effect, with morphine addiction. Several new drugs (heroin being a notorious example) had been introduced as nonaddictive, only to have addictiveness become apparent after use was widespread. Dilaudid, a semisynthetic opiate, had recently been brought onto the market with similar claims that, while it was a powerful analgesic, it did not seem to produce addiction; the literature contained conflicting reports on this point. Typically, a pharmaceutical company would distribute a promising new opiate, for which animals tests indicated useful therapeutic effects and adequate safety, to a network of clinicians and await their reports on how the drug performed. Addictiveness, as a side effect, was notoriously difficult to achieve consensus on.

At Michigan, Nathan Eddy grappled with the problem of demonstrating addictiveness with animal studies. The uncertainty about how to develop a test for addictiveness was reflected in a monograph describing the Desomorphine research: "The true nature of drug addiction is not known, and any attempt to determine and evaluate the presence of addictive liability in a compound of unknown potentiality must be based upon the limited extent of our present knowledge."[59] Eddy first tried continuous administration of Desomorphine in dogs, cats, and monkeys, checking for any signs of tolerance (reduced response to the accustomed dose), followed by abrupt cessation of administration and more observation. Initial results were inconclusive. Dogs failed to demonstrate tolerance or significant withdrawal signs. Cats showed tolerance to some drug effects but not others. Eddy compared continuous Desomorphine administration in monkeys to a similar regimen with morphine, and again pronounced evidence of tolerance and withdrawal phenomena was not forthcoming. This experimental method in some respects resembled clinical experience with new opiates: a drug was administered (the length of time determined by clinical indications), and physicians watched their patients for signs that they wanted increased doses and for the appearance of withdrawal symptoms following cessation of administration.

On the basis of his survey of the pharmacological literature, Robert Hatcher had suggested to the committee "a simple test which I believe to be available for determining whether a substance is actively habit-forming and capable of inducing addiction comparable to that to morphine and opium." Hatcher's "simple test" was based on his observation that "every drug which has found wide and effective use in preventing withdrawal symptoms during gradual withdrawal of morphine has been found to [be] itself actively habit-forming. . . . I suggest, therefore, that the substance to be tested be administered in suitable doses to addicts in place of the doses of morphine to which they are accustomed. If the substance fails to prevent the symptoms of morphine withdrawal, the substance will be found to be non-habit-forming or only slightly so."

Conversely, if the test drug, when substituted for morphine, suppressed the withdrawal syndrome, it was presumably addicting in the same manner as morphine. Such a procedure, Hatcher believed, would provide a reliable test that would avoid the length, complications, and expense previously associated with clinical studies of addictiveness—which, in any case, failed to provide conclusive results.[60]

Sollmann and Hatcher had both communicated with Treadway on the desirability of determining whether new opiates would act like morphine in suppressing the withdrawal syndrome in addicted individuals. Such studies would test the premise behind Hatcher's "simple test," and they required subjects already addicted and stabilized on a steady dose of morphine. As the official in charge of planning the PHS narcotics hospitals, which would house addicted federal prisoners and probationers as well as voluntary patients, Treadway was in a position to provide these. Treadway readily approved the type of study Hatcher had recommended. He saw advantages in the degree of experimental control afforded by incarceration and lengthy sentences.[61]

At the Fort Leavenworth Prison Annex, Himmelsbach faced the problem of testing compounds, including dihydrodesoxymorphine-D, for addictiveness. Having learned about tolerance from Nathan Eddy, and having at hand as research subjects prisoners who were actively addicted to morphine, Himmelsbach speculated that addictiveness, like other opiate effects such as analgesia, might display cross-tolerance—that is, that developing tolerance to certain drugs conferred tolerance to some similar drugs as well. This idea embodied Hatcher's notion that substituting a test drug for morphine in an addicted subject would reveal whether the test drug was also addictive. To operationalize the assay of addiction as a toxic effect, Himmelsbach developed a scale for measuring the severity of the withdrawal syndrome. Drawing analogically on the four-stage system (one to four plus signs) used to measure Wassermann reactions, Himmelsbach grouped individual signs and symptoms of withdrawal into four categories, which he characterized as mild (+, e.g., lacrimation), moderate (++, e.g., muscle tremor), marked (+++, e.g., insomnia), and severe (++++, e.g., vomiting).[62] Examining cases of morphine withdrawal, he plotted onset of symptoms against time and correlated overall severity with doses of morphine the addict had been taking. Plotting the results produced a graphic representation of the withdrawal syndrome.

Next Himmelsbach attempted the substitution study. He stabilized a few addicts on doses of morphine sufficient to stave off withdrawal symptoms, and then substituted Desomorphine for the morphine. Like morphine, Desomorphine prevented onset of withdrawal symptoms. Approximately ten days later, when Himmelsbach abruptly ceased administration of Desomorphine, all subjects quickly developed full-blown withdrawal syndromes according to the timetable and dose-related severity Himmelsbach had plotted. As a precise profile of the withdrawal syndrome emerged, and as Desomorphine substituted effectively for morphine in preventing withdrawal, Himmelsbach concluded that the test drug was addictive in the same manner as morphine.

The committee now had to reconcile contradictory results from Himmelsbach's clinical tests and Eddy's animal studies. If Himmelsbach's results were accurate, then Desomorphine, contrary to early hopes, was not the desired nonaddicting analgesic. In light of the psychiatric theory that chronic addicts like those imprisoned at Leavenworth possessed a psychiatric deficiency that differentiated them from "normals," committee members hypothesized that different results might be obtained with patients at the Massachusetts tuberculosis and cancer clinics.[63]

Himmelsbach traveled there in the spring of 1935 and followed the procedure Eddy had followed with his monkeys: he gave Desomorphine to patients who previously had had no opiates. However, Himmelsbach had recognized that Desomorphine's duration of action was substantially less than that of morphine; its effects wore off in two to three hours instead of four to six hours. With the addicts at Leavenworth, and now with the Massachusetts patients, he adjusted the frequency of administration to ensure a continuous effective dose. He argued that Eddy had failed to demonstrate addiction with the drug in monkeys because he had administered it on the same dosing schedule as he had the morphine. On

April 14, 1935, Small, Edmunds, and Eddy met at the Pondville Hospital in Massachusetts. Himmelsbach stopped the administration of Desomorphine to his research subjects, and the classic signs of withdrawal ensued. These symptoms disappeared when the patients were again given Desomorphine. Himmelsbach argued that the burden of proof that his results were unsound now lay with Eddy's group at Michigan.[64] Edmunds, Eddy, and Small all retained doubts, but Treadway supported Himmelsbach and suspended further tests on the grounds that its addictiveness made the drug unacceptable as a medication.[65]

Meanwhile, as Himmelsbach was preparing for the Massachusetts demonstration, Eddy started a new round of Desomorphine experiments that duplicated Himmelsbach's morphine substitution procedure with monkeys. As the results consistently confirmed Himmelsbach's findings, Eddy and Himmelsbach co-authored a 1936 article describing the method and reporting the finding that Desomorphine was addictive.[66] Himmelsbach and his colleagues at Leavenworth had also published the results of a morphine substitution study with Dilaudid that showed that drug was also addictive, countering numerous clinical reports to the contrary.[67]

Himmelsbach's findings were challenged by reports from European clinicians that Desomorphine did not produce addiction among patients to whom they administered it. Chemists at Hoffmann–La Roche, a pharmaceutical firm in Basel, had synthesized dihydrodesoxymorphine some ten years earlier but had not investigated the compound. When the American group reported its strong pharmacological activity, the European firm quickly took the compound off the shelf and arranged for a number of clinicians to test it with their patients. Lyndon Small returned from a visit to Basel with a thick sheaf of European clinical reports reflecting a consensus among these physicians that Desomorphine was preferable to morphine as an analgesic because it did not produce sleepiness or mental confusion; they also asserted that addictiveness was not a problem.[68] Himmelsbach read the reports and recognized why addictiveness had not been detected: the clinical situations had not provided the right conditions for addiction to occur or, when it did occur, practitioners who lacked precise knowledge of its signs failed to recognize it. Case by case, Himmelsbach noted factors that had not been accounted for or sufficiently controlled. Data on dosage and rhythm and duration of administration had not been recorded. When the drug was given for postoperative pain, the period of administration was simply too short for addiction to develop. Desomorphine was short-acting, and too long an interval between doses would result in intermittent rather than continuous effective dosage. Patients already addicted to morphine developed withdrawal syndromes when given Desomorphine because the dose of the latter did not correspond to the accustomed morphine dose.[69]

As Himmelsbach's findings continued to stand up to scrutiny, the morphine substitution method, with refinements, became the standard assay method for addictiveness. The Americans' success in formulating this standard depended in part on methodological ingenuity in devising the morphine substitution method. However, the availability of subjects who were already addicted to morphine was also crucial. Patients receiving analgesics for medical indications were not suitable subjects because too many factors in the clinical situation clouded the addiction picture; doses and duration of administration were determined by clinical rather than experimental necessity. The availability of addicted prisoners made it possible to create virtually identical study conditions in the animal laboratory and the research ward. Lexington inmates—federal prisoners addicted to opiates but otherwise healthy—thus represented the "right tools for the job" of creating an addictiveness assay.[70]

The negotiations over the meaning of Himmelsbach's finding that Desomorphine was addictive illustrate the tensions between laboratory scientists and clinicians in the 1930s as medical reformers sought to establish a scientific basis for evaluation of treatments.[71] Laboratory scientists insisted on the importance of standardization and control to isolate variables and ensure uniformity of results. Clinicians, however, faced the human and pathological uniqueness of each patient and argued that knowledge born of long experience, not technical data, formed the foundation for understanding the condition of the patient; moreover, they worked in settings that rarely had the equipment or trained personnel necessary for laboratory work. The ambiguous claims of the late 1920s regarding Dilaudid's addictiveness reflect the common pharmaceutical company practice of sending samples of promising compounds to clinicians who often reported results based on administering them in the course of their normal practice rather than in controlled studies. Himmelsbach's commentary on the equally ambiguous European reports regarding Desomorphine reveal the kinds of clinical conditions that made systematic observation of specific treatments difficult in hospital settings.

Harry Marks has shown that clinical researchers in the 1930s urged cooperative research structures and individual clinical expertise grounded in proper understanding of science as the basis for determining the efficacy of treatments; nevertheless, he argues, how to translate laboratory findings into therapeutic tools remained contentious.[72] The CDA-sponsored research was a cooperative project with clinical goals that linked four sites: the chemical laboratory at the University of Virginia; the pharmacology laboratory at the University of Michigan; the prison research ward at Leavenworth; and the cancer ward at the Pondville Hospital. But when Himmelsbach's results contradicted findings from the animal laboratory, how to resolve the conflict was by no means clear. Himmelsbach faced the problem of convincing both the animal pharmacologists and the clinicians of the validity of his results. This he did first, with the demonstration at Pondville and, second, through instructing Nathan Eddy in his methods, which held up to continued scrutiny. Thus, knowledge did not flow from the laboratory to the hospital ward as an idealized model of knowledge production might suggest. Rather, the prison research ward, where addicted, but otherwise healthy, human subjects could be studied with levels of control comparable to those achievable in an animal laboratory, proved a crucial site for developing the addictiveness assay.

Despite the elusiveness of the nonaddicting analgesic, gains were sufficient to keep the work going. When the Bureau of Social Hygiene withdrew its funding, the Rockefeller Foundation picked up the funding for the Committee on Drug Addiction in 1932, and it renewed it in 1936. The shift from Bureau of Social Hygiene to Rockefeller Foundation sponsorship illustrates how fully the work met current standards for innovative and worthwhile biomedical research. When the Rockefeller Foundation announced plans to cease funding the work, the National Institute of Health brought the chemical and pharmacological components to its Washington, D.C., site in 1939.[73] Among the benefits of this work, Eddy listed contributions to the understanding of relationships between structure and activity in opiate pharmacology, improved understanding of drug dependence, and the ongoing progress toward safer analgesics.[74] Many publications in the chemical and pharmacological literature, as well as several important monographs, had resulted. The first edition of Louis Goodman and Alfred Gilman's textbook *The Pharmacological Basis of Therapeutics* cited Small, Eddy, Himmelsbach, and their co-workers extensively in the discussion of opiates and hailed the Committee on Drug Addiction project as a promising source of new and improved opiate medications.[75]

Pharmaceutical companies also realized significant benefits, despite the meager returns in marketable drugs. First, they had gained highly trained personnel. Second, Small and Eddy's systematic approach to a class of potential compounds revealed dead ends that companies might otherwise have pursued at their own expense. Less tangible benefits derived from association with a prestigious academic research project. Small and Eddy's receipt of the first Annual Scientific Award of the American Pharmaceutical Manufacturers Association in 1939 reflected the perceived value of their work. That Harry Anslinger presented the award and, in the presentation speech, linked their scientific work to international drug control efforts further emphasizes the structural links between the search for a nonaddicting analgesic and federal drug policy.[76]

Finally, one marketable product, Metopon (methyldihydromorphinone), a compound developed by Small, did emerge from the work in the 1930s. Its therapeutic use was restricted to the treatment of chronic pain in cases of advanced cancer, and it was withdrawn from the market in the early 1950s. However, its acceptance as a medication illustrates that for scientists engaged in drug development, addiction had become a measurable side effect subject to the same calculus of risks, benefits, and marketability as any other drug effect. Metopon was an effective analgesic with less respiratory depression and emetic effect than morphine. In assessing its addiction potential, Himmelsbach compared it with three other compounds, including Desomorphine. His quantitative assessment of the intensity of withdrawal syndromes enabled him to show that Metopon, although addictive, was demonstrably less so than morphine or Desomorphine. When Metopon was given to subjects who had been stabilized on morphine, a few mild withdrawal symptoms did occur. Thus, Metopon did not substitute completely for morphine in satisfying the preexisting addiction. Conversely, when the Metopon was stopped, the ensuing withdrawal syndromes were milder and shorter than those associated with morphine.[77]

Extensive clinical tests in Massachusetts again compared Metopon to morphine and Desomorphine. These tests reconfirmed that Desomorphine lacked significant advantages over morphine. In treatment of pain in cancer patients and acute trauma patients, Metopon provided effective analgesia at lower doses than morphine, with fewer side effects, including less tendency than morphine to produce mental clouding. A tendency to produce respiratory crises when used as a calmative before inhaled anesthesia contraindicated its use in these situations. Although morphine clearly remained the best available general analgesic, Metopon's production of milder addiction and the quick fading of tolerance to the drug during intervals of abstinence recommended it for use in cases of chronic pain.[78]

Metopon had shown that addiction potential could vary independently from analgesic potency. For those interested in drug development, recognition of the relative nature of addictiveness held out the promise of less addictive drugs even if the goal of eliminating addictiveness remained beyond reach. For pharmacologists and the chemists working with them, the ability to manipulate analgesia and addictiveness separately supported the assumption that pharmacological effects were dependent on molecular structure. Further research to elucidate relationships between structure and activity was clearly warranted. In both practical and scientific terms, the research plan adopted by the Committee on Drug Addiction yielded important payoffs.

At the same time, the merely incremental improvement that Metopon offered over analgesics already in use, including morphine, fell far short of the real goal of the committee: an analgesic that would produce powerful analgesia with no addiction. In this way, Metopon was a typical product of the methods of classical pharmacology. In a 1928 overview of the Committee on Drug Addiction's work, Alfred Newton Richards had seen the likelihood of just such an outcome: "It seems almost self-evident that it will be impossible to discover any substance having all the desired actions of morphine and which will not produce addiction."[79] Richards acknowledged that organic chemists' ability to produce vast numbers of structural variations of particular compounds had raised great hopes. "However," he continued, "since 1850, only a few active basic substances have been found with few gains for therapeutics, but a series of variations of basic substances having equal or less value, and differing only in unessential grouping of molecules have been introduced." Richards recognized, but did not think well of, the marketing potential in such a situation: "No new actions possible [*sic*] from such compounds, but new actions must be attributed to them in order to sell them." Thus, slight changes in structure could yield novel, patentable compounds that could be marketed without infringing on rivals' patents but could compete directly with other products. Richards interpreted this situation in negative terms: "Despite efforts of discoverer, manufacturer and physician to limit market to one, or at most, two representatives of each group, competition results in a flood of substances having equal and similar actions. Despite the lesson, investigators and manufacturers continue to try to find variations of similarly acting substances which will exert effects unlike those of the group."

Richards, though, was speaking as a scientist, not as a manufacturer or a physician. While patent holders for successful drugs certainly resented the entry of rival compounds into their markets, drug manufacturers aggressively exploited the opportunities to capitalize on successful therapies by developing new compounds that could compete directly with existing ones, and this form of competition has resulted in the marketing of similar medications ever since. And physicians, aware of the idiosyncratic ways in which patients may react to both the therapeutic and unintended effects of drugs, can benefit from an array of similar medications to choose from in clinical practice. Even "investigators," as Richards noted, often profited from the situation, as they were able to secure pharmaceutical company support (whether in the form of salaries as employees, or in the form of grants or fellowships in the academic sector) for research in organic chemistry and pharmacology. Reflecting the divide between physiological and drug-development pharmacology, Richards lamented the lack of truly new findings in the latter: "Physicians as well as chemists have *no clear idea* upon what specific activity certain classes of substances rest, and . . . only by analogy with other groups, new substances are produced, and the finders are delighted because they have therapeutic properties, unmindful of the fact that only unessential changes in constitution have been introduced." He cited heroin as an example of this process.[80]

Richards's ruminations reflected the ongoing tensions about what constituted basic science in pharmacology. In the 1930s, Lyndon Small and Nathan Eddy could argue that the communication between their two laboratories, comparing effects of new compounds to the structural changes that distinguished them from other compounds, produced meaningful findings on structure-activity relationships, and their work had been framed at the outset as having the potential to improve understanding of addiction. But for Richards, the failure to seek the mechanisms whereby drugs achieved their effects dimmed the scientific luster of the work.

In time, the search for a nonaddicting analgesic would seem like applied science that had failed to deliver on the promise of improving understanding of addiction.[81] However, Eddy, Small, and Himmelsbach concluded the testing of Desomorphine and Metopon just as the most dramatic breakthroughs in therapeutics were about to burst on the scene. By 1939, the year that Small and Eddy moved to the NIH to continue their work in the same institutional setting, the sulfa drugs, the first effective treatment of infections, had been available since 1936, and penicillin's power to cure infections had just been established. The triumphant project of scaling up penicillin production to meet the needs of a world at war was about to begin. However contested the value of drug development as basic science, its value in the clinic and the marketplace was clearly established. Nevertheless, as Richards had noted, heroin resulted from the kind of competition that drove drug developers to spin out hundreds of similar compounds with slightly different constellations of effects. As Lawrence Dunham also recognized, the profusion of new medications to treat human disease was also the source of new psychoactive drugs, providing new menus of means to alter human consciousness.

Conclusion

The work of the Committee on Drug Addiction merits analysis on at least two levels: at the level of a biomedical discipline undergoing institutional growth in the context of American medical reform and large-scale organization in American science; and at the level of the interplay among social reform currents, federal policy, and scientific entrepreneurship in the period between the wars. The significance of the committee's work lies in part in its success in capturing significant resources and deploying them effectively in a problem area important to pharmacologists and others. However, the committee's ability to pose a drug-development solution to the social problem of addiction reflects a particular constellation of private and public sector actors and settings, each with its own combination of larger aims and immediate objectives. The success of the Committee on Drug Addiction lay in good part in the way its needs fitted into a neat mosaic with the concerns and goals of other groups involved in addiction research and policy.

The Bureau of Social Hygiene handed its addiction work over to the NRC in 1929, a year after the Rockefeller Foundation shifted its funding policy away from capital grants to universities in favor of grants for specific projects, especially ones cutting across disciplinary boundaries.[82] This meant a shift from the idea of supporting "excellence," consistent with Katharine Bement Davis's opportunistic funding approach, to a more strategic sense of what research should be supported. The Committee on Drug Addiction was also one of numerous NRC efforts in the late 1920s and 1930s to bring scientists together to work cooperatively on problems that straddled disciplinary borders.[83] In part because the committee reached early and stable consensus on Hunt's research program, it avoided some problems that affected other, similar efforts. Like the Committee for Research on Problems of Sex (CRPS), the Committee on Drug Addiction owed its existence to concern for a social problem aroused via the work of the Rockefeller-funded Bureau of Social Hygiene. Here, too, the resources shunted to the NRC were captured by a group of scientists focused on discipline building and crucial internalist problems.[84] Unlike CRPS, however, the Committee on Drug Addiction did not argue over cognitive models on which to base its research. The effort was not to create new kinds of science but to catch up with institutional structures and research methods already known to be useful in the discovery of new medications.

Unlike NRC-sponsored work in biophysics, the Committee on Drug Addiction's project did not call for the application of instruments developed in one discipline to research materials in another; nor did it train scientists educated in one discipline in the methods of another. The project's university-based scientists worked cooperatively; they and the committee met frequently to discuss results and tactics, especially the growing body of correlations between pharmacological activity and molecular structure. Correlations, by their nature, allowed each group to work in its own laboratory on problems within its own defined discipline; these disciplines, organic chemistry and pharmacology, fell neatly within existing university department structures. Each group was able to publish in the journals of its own discipline (the chemists in the *Journal of the American Chemical Society* and the pharmacologists in the *Journal of Pharmacology and Experimental Therapeutics*). Each group's work (development of novel compounds and elucidation of their chemical structure; assays of pharmacological effects on various physiological systems) produced meaningful findings within established disciplinary channels of communication. The link consisted of test compounds, which moved among the chemists, pharmacologists, and clinical researchers. Although the Desomorphine episode showed that conflicts over meanings of results could occur, cross-disciplinary conflicts among scientists over how the disciplines should interact or what problems to study did not arise.

The search for a nonaddicting analgesic emerged as a project typical of the "classical pharmacology" of the 1930s, in which bioassay of compounds revealed therapeutic or toxic effects and their related dose ranges. Such research proved skillful at developing certain forms of new medication, as Richards had observed—most notably those that displayed comparative advantage over existing medications rather than revealing bold new directions for therapeutic research. Thus, toxic effects could be muted, or desired effects could be amplified; similarly, multiple therapeutic effects associated with a compound could be separated so as to produce a range of compounds, each with a distinct beneficial effect. Metopon's marginal improvement over Desomorphine illustrates these characteristics of pharmacological research in the classic era.

The committee's work also parallels wider institutional trends in academic pharmacology in the United States in the 1920s and 1930s, when university departments and pharmaceutical firms were forging important links. Pharmacology departments received support and were helped in developing specific research tasks in the period of post-Flexnerian reform, when creating new research missions was a vital disciplinary activity. Pharmaceutical firms, which were both building their own research and development facilities and striving to distance themselves from the stigma associated with purveyors of nostrums, valued the prestige of university research to back up their product claims in the medical marketplace.

The goal of developing improved opiate analgesics also proved consistent with professional concerns of the American Medical Association and reform-minded and professionalizing physicians. Policies that increased physicians' control over who received opiates were welcome adjuncts to the AMA's own efforts to reduce indications for opiates to an irreducible minimum. A project to develop a nonaddicting opiate analgesic clearly fitted well with the AMA reform agenda.

In these ways, the NRC-funded search for the nonaddicting opiate fitted into the larger pattern of institutional development of American science and medicine in the 1930s. But in several respects, the project was unusual, and these uncharacteristic factors were also important to the successes the researchers achieved. Chief among these was the illegality of the drugs involved, which made government involvement in the committee's work inevitable.

Because morphine and related compounds could not, by law, be possessed unless prescribed by a physician, special arrangements had to be made to give scientists access to the amounts they needed for research. Small, Eddy, and others were made $1-a-year employees of the Public Health Service so that they could be authorized to handle the drugs. Researchers not involved in the project faced bureaucratic hurdles in obtaining these drugs for research. In practice, this meant that the committee would send small amounts of drugs to scientists they knew and respected as long as their projects did not overlap with the committee's work.[85] The special legal status of the substances discouraged some researchers who might otherwise have undertaken work similar to Small's and Eddy's. The project scientists thus had a measure of protection from competition.

More important than the legal status of the drugs was the legal status of opiate addicts in the United States at this moment in American history. As the legislation calling for construction of the PHS narcotics hospitals moved through Congress, White already envisaged cooperation between the committee and hospital staff. When addicts were deemed useful as clinical subjects, the institutional groundwork to gain access to addicted prisoner-patients had already been laid. A frequent source of difficulty in drug development in the 1930s was the identification of appropriate clinical test settings and methods. Himmelsbach later described how, in the research ward at Lexington, he achieved virtually complete experimental control over the conditions of his research—virtually, because even with a locked ward, carefully screened personnel, and monitored food preparation, an especially wily addict might succeed in smuggling in drugs. One prisoner kept a tiny stash of morphine in a hollow area created by oral surgery and confounded Himmelsbach when withdrawal signs failed to occur on schedule—until the subterfuge was discovered.[86] This degree of control made it possible to isolate the phenomenon of addiction from the web of complicating factors that clinicians encountered in patients. As a result, he and Eddy were able to devise closely comparable protocols for human and animal tests for addictiveness.

Opiate addiction, in the 1920s and 1930s, provided a nexus around which various social reform and scientific constituencies could develop agendas, marshal and allocate resources, and advance their own work. These constituencies included social reformers, biomedical scientists, physicians, policy makers, public health officials, and enforcement officials. To varying degrees, they focused on addiction as a primary concern or, like Reid Hunt and the NRC's Committee on Drug Addiction, they responded opportunistically to a chance to pursue disciplinary agendas. In either case, they approached opiate addiction as an important and real problem, to which they hoped to offer a meaningful solution. The problem areas staked out by each constituency and the proffered solutions, viewed ecologically, reveal a pattern of complementary or mutually reinforcing needs.

At a time when American science was becoming a major enterprise, the case of laboratory research on addiction illustrates how the institutional structures of science grow through interactions among local contexts, individuals' agendas and entrepreneurial efforts, and both the programmatic and the methodological demands of individual disciplines. It also points up the need to examine scientific efforts in broader contexts of social problems and related policy. Social concerns about addiction, close links between NRC scientists and federal officials, and an unquestioning approval of federal narcotics policy by the scientists involved all helped shape research concerns, which played out in choice of research questions and in development of experimental design.

The interests of pharmacologists in the classical era proved closely compatible with the structures of drug policy in its own "classic" era. The period prior to the passage of the Harrison Narcotic Act in 1914 had been characterized by heterodoxy in understandings of addiction and how to treat it; moreover, the stigma associated with opiate addiction was muted by sympathetic portraits of respectable individuals drawn helplessly into enslaving habits through their own or a physician's carelessness. Since 1914, implementation of the Harrison Act had criminalized addiction. A legislative and policy approach that called for restricting imports of opiates to the amounts needed for legitimate medical supplies and segregating addicts from the larger society replaced earlier meliorist approaches. The Bureau of Social Hygiene's commitment to the League of Nations treaty-making process and the participation of the NRC's Committee on Drug Addiction in that process illustrate how the pharmacologists' objectives were seen as consistent with the policy thrusts of both the federal government and the Rockefeller philanthropies. Developing a nonaddicting analgesic that would obviate any need for morphine in the pharmacopoeia fitted well with the goal of restricting entry of unauthorized opiates into the United States. Thus, the scientific objectives of the pharmacologists implicitly supported the supply-side drug policy approach that was developed in this period and has characterized American drug policy ever since.

4

Constructing the Addict Career

If the search for a nonaddicting analgesic under the Bureau of Social Hygiene's sponsorship was moderately successful, its efforts to promote a psychiatric understanding of addiction met with frustration and failure. BSH-sponsored research at the Philadelphia General Hospital was designed as a comprehensive study of the clinical material provided by the hospital's narcotics wards. It included psychiatric and psychological interviews with patients and an array of physiological examinations. The Committee on Drug Addictions had trouble both retaining a psychologist and obtaining adequate reports from the psychiatrist engaged to work on the study. More fundamentally, however, the attempts to construct a psychiatric understanding of the opiate addict and to provide a form of counsel that would support continued abstinence after hospital detoxification failed because psychiatrists in the 1920s lacked the therapeutic resources for successful interventions in the lives of opiate addicts. Despite the existence for some decades of sanitaria dedicated to treating inebriety, and despite the flurry of experimentalism that characterized addiction treatment in the 1910s, opiate addiction remained a poorly understood condition for which no clearly recognized effective treatment existed.

Several factors inhibited progress toward better understanding of addiction in the 1920s. As the pattern of opiate addiction shifted from the late nineteenth century to the 1920s, the iatrogenically addicted female patient was replaced by the typically male pleasure seeker, who often rejected medical advice and affronted physicians' middle-class values. This type of addict presented a new kind of puzzle for clinicians, who drew on a range of current diagnostic influences in psychiatry to explain what often seemed willfully obstinate behavior. These diagnostic influences included the fluid concept of psychopathy, degeneracy theory, eugenics, and the idea that feeblemindedness explained much deviant and criminal behavior, and they were conveyed through diagnostic labels such as "constitutional inferior," "defective personality type," and "psychopathic individual." They had in common an essentialist basis that assumed that the disturbed behavior that betokened mental illness arose from fundamental defects of personality.

By the mid 1920s, the opiate withdrawal syndrome was not only well recognized but for some time it had also been the diagnostic indicator of last resort when opiate addiction was suspected. This implicitly equated the withdrawal syndrome with addiction itself. Addicts' demands for treatment, although couched in terms of a desire to shed the habit, were often a cry for help in getting through the rigors of withdrawal. Thus, from both the clinician's point of view and the addict's, the withdrawal syndrome occupied a primal space in the meaning of addiction. Most treatment regimens consisted of medicating or palliating the period of withdrawal. In time, stubborn patterns of repeated relapse would indicate the need for some longer-term form of therapy subsequently, but in the 1920s, treatment focused on the withdrawal syndrome. The tendency to relapse was increasingly seen as evidence of the addict's preexisting susceptibility to addiction rather than a failure of the treatment method, raising mounting doubts about any possibility of cure.

As discouragement set in and treatment options disappeared, the possibilities of clinical research were being all but eliminated. A generation of physicians received virtually no education in the treatment of opiate addiction and were unlikely to encounter it in their medical practices, if only because they failed to recognize it in patients. In the absence of a dynamic engagement with opiate addiction as a clinical problem, psychiatric explanations of addiction that were consolidated in the 1920s held sway until the 1970s. Such research as did occur (most of it at the Public Health Service Narcotic Hospital at Lexington, Kentucky, which opened in 1935) was based on the assumption that a psychopathic or sociopathic personality type underlay addiction.

The Reformer's Nemesis

A psychiatrist's interviews with addicted patients in the narcotics wards of Philadelphia General Hospital in the course of research funded by the Bureau of Social Hygiene reveal a great deal about addict life in the 1910s and 1920s. They also illustrate the kind of conundrum that the new type of addict presented to physicians who struggled to guide their patients to an abstinent life. Their efforts foundered on the intractability and willful misbehavior of patients like Michael Heider.

A high-minded Progressive reformer's nightmare, Heider drifted in and out of low-paying, short-term jobs and of affairs with women, with no thought of marriage even when one of them became pregnant.[1] He had become addicted to heroin from age 17 in 1913; he gambled; he got into fights at dance halls that had to be broken up by the police. By 1926, when he came to Philadelphia General Hospital, he represented an increasingly well recognized type of intractable addict. The intractability of his addiction was evidenced by his return to the narcotic ward in 1927 and again in 1928.

These three years afford a rare window into the lives of opiate addicts in the American Northeast in the first quarter of the twentieth century. Under the auspices of the Bureau of Social Hygiene, addicted patients' bodies and psyches were probed in an effort to better understand addiction and how to treat it. The internist Dr. Arthur B. Light led a team that examined many aspects of these subjects' physiology; a psychiatrist, Roy B. Richardson, interviewed them repeatedly and at length; and psychologists assessed their mental status. The records of this research effort necessarily limit our view of these patients to the responses elicited by the researchers, but, thanks to what Elizabeth Lunbeck calls "the newly inductive nature" of psychiatry in this period, its concern with everyday life, and the presence of a stenographer who captured long passages of patients' responses, the records are rich with detail about these addicts' lives and experiences.[2]

Nevertheless, there are many limits to the transparency of these records. Some of these lie in the subjects' assessments of their hospital experience. As addicts seeking detoxification, they wanted medication of their withdrawal to decrease its agonies, and some sincerely hoped that their treatment would lead to a real, lasting cure. The physician researchers often admitted patients without charge in exchange for their willingness to be studied, and these subjects shaped their behavior and responses to some extent in an effort to conform to their perceptions of the researchers' expectations. Some were expelled untreated for failure to cooperate with the research team. Even if they had come just for treatment (and allowed themselves to be studied after admission), many responded to their doctors' admonitions and advice with an earnest resolve that they may not have truly felt. The administration of morphine in the first hospital days, to stabilize them and allow preparation for treatment, was an incentive for some addicts who had run out of drugs and become desperate. Some expressed concern about the uses of the information they yielded—would it be passed to the police? would contacts they named be sought out?—and modulated their responses accordingly. Often, though, assurances that the interviews were intended to benefit them and others in similar plights, and the sheer constancy of the bedside visits and questions, prompted matter-of-fact answers. Finally, the entire hospital stay involved a transition from morphine administration through the rigors of withdrawal and into several days of bed rest, regular meals, and palliative medication after withdrawal and before discharge. As Light and his colleagues noted, addicts' behavior varied markedly as they moved through these stages.[3]

By the 1920s, addicts had long had a reputation among physicians, as well as the general public, for telling lies to obtain drugs.[4] Even aside from this motivation, addicts had many reasons to be less than truthful in a situation like that at Philadelphia General Hospital. The stigma of addiction and the danger of arrest created obvious motives to alter identifying information like names and addresses. Moreover, by the time they came to the hospital, many addicts had become adept at living double lives. They had learned how to play to expectations about drugs and drug users. In the hospital, they were expected to play the role of patient and research subject, and many learned to balance these expectations against their own needs: they might be seeking a respite from a burdensome habit or hoping to convince concerned family members that they were sincerely trying to end their addiction.

In the context of Progressive Era immigration, and as addiction increasingly was concentrated in the social contexts of amusements, vice, and crime, drug use was just one of several reasons for concealing the truth, and addicts might be familiar with numerous contexts where aspects of identity were manipulated. Many immigrant families changed their surnames when they came to the United States, either to make themselves seem more American or when immigration or school officials misrecorded names or altered their spelling to correspond to English pronunciation. People who worked in the theater world commonly renamed themselves; so did prostitutes working in brothels.[5] Some taxi dancers managed the potential stigma associated with the downward status shifts involved in entering the dance hall world by manufacturing not only names but new identities.[6] Criminals had obvious reasons to hide their identities, and by the late 1920s, addiction itself had become a crime. In this period, a synergistic relationship between the world of vice and the increasing repression of that world (through such measures as the Harrison Narcotic Act, alcohol prohibition, and the suppression of prostitution) increased the motivation to lie to authority, including medical authority. Finally, for many addicts from working-class neighborhoods in or near urban Tenderloins, a hospital stay may have been an unusual confrontation with middle-class expectations of respectability, narrowly defined in terms of gender roles, sexuality and sexual behavior, thrift, future orientation, and decency. Thus, the responses of addict-patient-subjects to the probing questions of a psychiatrist in a hospital-based research study must be interpreted with care.

Agreeing to be a research subject at Philadelphia General Hospital's narcotics ward meant signing on for a rigorous regimen. When Michael Heider was admitted for his first visit on June 10, 1926, he was stripped of the clothes he wore and whatever he had brought with him. On the ward, where he wore hospital-issue pajamas, robe, and slippers, he was dosed with morphine. Light took a history of his drug use, which elicited three previous attempts at treatment, all in Michael's home city of New York. Light also conducted a physical examination, which revealed a poorly nourished and poorly developed adult male with dark brown skin. Michael's upper arms (presumably above the sleeve line, where they would be hidden from casual view), chest, abdomen, and thighs were scarred from repeated hypodermic injection; in some places, the skin had been covered over with scar tissue. Several of Michael's teeth were missing and those that remained were "blackened stumps." A Wassermann test indicated he had never been infected with syphilis. Richardson also visited Michael on his first day in the hospital. Besides again eliciting information about Michael's drug experiences, Richardson asked about his attitudes to drugs, his family life, his relatives, his childhood, his school life, his occupational history, his tastes in amusement, and his relationships with women.

On June 12 through 15, still receiving regular doses of morphine, Michael underwent a series of physiological tests. Blood was frequently drawn; respiration and blood pressure were repeatedly measured. On June 12, he ran a half mile in four minutes and fifteen seconds, and then his pulse was taken at fifteen-second intervals for five minutes. His intraocular pressure was measured. He ran up two flights of stairs, and his pulse was taken again. His blood chemistry was checked; an electrocardiogram was taken; and he underwent liver and kidney function tests. Richardson interviewed him again and conducted a neurological examination, and a psychologist did a mental assessment.

On June 16, the morphine injections ceased, and Michael began to go into unpalliated withdrawal. Richardson visited him frequently to ask how he was doing. The second day after withdrawal of the drug was the hardest, Michael reported; on this day, the various tests, including the run up two flights of stairs, were repeated. Richardson acceded to Michael's request that he not be interviewed that day. Late in the day, Michael was again given morphine, and the relief of his withdrawal symptoms was precisely noted and timed.

On June 19, the morphine was stopped again, and Michael began the standard treatment regimen: medication with hyoscine during withdrawal, along with cathartics for constipation, luminal (a barbiturate) to help him sleep, and small doses of strychnine as a stimulant. He spent two days in a confused hyoscine-induced delirium, receiving his last dose of hyoscine on June 21. The next day, Richardson administered a sensorium, a battery of questions to test simple reasoning ability, knowledge of current events and of rules of capitalization and punctuation, and a variety of other cognitive tasks. On June 23, Michael reported that he could eat a meal and keep it down for the first time since the morphine had been withdrawn.

The next day, despite a stated wish to go home and a failed attempt to bribe a staff physician to let him out, he repeated the round of physiological and stress tests for a third time, this time in the abstinent state. Richardson's interview on this day focused on Michael's prospects of continued abstinence. The psychiatrist probed for regrets about the past and future plans. As the withdrawal symptoms subsided, Michael made light of the bribe attempt: "I was going to give him $15.00 to let me go, but I wouldn't give him a nickel today and tomorrow I wouldn't give him two cents."

On June 26, Richardson noted that Michael's gruff and irritable manner during withdrawal had given way to a cheerful optimism. Michael professed a need "to make good this time" and promised to throw away a half ounce of heroin he had saved as soon as he got home, although his earlier histories indicated a pattern of resumed use almost immediately after treatment. Finally, on June 28, after a mental examination, Michael was discharged, pronounced "Relieved."

The results of tests like these with dozens of subjects formed the basis of Light and Torrance's physiological conclusions about opiate addiction: that morphine or heroin caused few observable enduring changes in the addict's physiology; that those they noted were benign; and that the withdrawal syndrome was not life-threatening—and, thus, by implication, could be managed without medication.[7]

Michael's responses to Richardson's questions indicate that he was born in the United States of parents who had immigrated from Austria, the youngest of eight children who survived early childhood. He attended school through the eighth grade, performing indifferently. Finishing school at age 14, he spent two years living at home and not working; when he did work as an adult, it was usually as a machinist operating a printing press. He was introduced to opiates in about 1913—before passage of the Harrison Act—when a friend offered him powdered heroin to "blow," or sniff. He declined initially: "Just looking at it made me throw up." But a week later, he tried it, began blowing it occasionally on weekends, and gradually began taking it on Mondays too, then Tuesdays, until he was using it every day. One day when he did not take any drugs and began to get sick, he realized that he had become addicted. After several years, he began injecting the heroin instead of sniffing it, because he preferred the drug sensation with this route of administration.

Michael found that the drug habit interfered with his work; in any case, his employment was irregular. When his mother died, leaving $600, which he appropriated on the grounds that he was the only sibling living with her at the end of her life, he quit his job and spent the money enjoying himself at his favorite pursuits, which included going to the fights, taking women to dance halls, playing pool, and gambling. For a while, Michael obtained his drug from the municipal clinic established in New York City to treat addicts suddenly bereft of drugs following the implementation of the Harrison Act. He had sought treatment for his habit a number of times before coming to the Philadelphia General Hospital; he had also undergone the hyoscine treatment at King's Hospital in New York.

Michael's exchanges with Richardson reveal the differences in values and attitude between the patient and his professional questioner. Richardson sought to draw out relevant aspects of Michael's history, but many of Michael's responses displayed the existential aplomb of a gangster in a Godard film—an individual who refuses to get caught in the analytical framework through which the larger society seeks to judge him. Asked why he took drugs, Michael replied, "I don't know why I take it, just wanted to try it." Probing his connection with a drug-using associate, the psychiatrist asked, "How did you get into a conversation with him?" Michael replied, "How does anybody get into conversation? That's just the way this happened." And when Richardson tried to elicit a sexual history, Michael said, "I don't remember why I liked my first girls. Why do you like girls, doctor? Same as I do, same with me. I like to just go with a girl because I love a girl." In fact, he loved a lot of girls, as he then recounted a long string of casual sexual relationships.

Michael Heider sought treatment repeatedly because his drug habit had become a burden. He assured the doctor that he would make good this time, but he also acknowledged that he would typically start using heroin again the day after leaving a hospital. Heider typifies in many respects the young, male, urban addict who represented the new pattern of addiction that emerged in the United States from about 1900. He began using heroin socially, exposed to it in the context of masculine working-class entertainment sites. He drifted between legitimate employment as a printing shop machinist and idle periods, which he spent attending fights and spending time with women. His siblings included brothers with respectable jobs and a brother who owned a speakeasy. He had numerous scrapes with the police—for example, as the result of getting into fights at dance halls. His attitude toward Richardson, which varied between easy sociability when he was comfortable, mild irritation with some of Richardson's probing questions, and a pulling into himself while he went through the by-then familiar ordeal of withdrawal, suggest that he was polite on his own terms, willing to cooperate when it was easy and convenient for him to do so, but lost patience with questions that seemed to him meaningless or unreasonable. His pattern of frequent cures and immediate relapse, typically using heroin on the same day he left the hospital, suggest that his cures were more a tactic in the management of addiction than a commitment to ending his drug use permanently. Nevertheless, his statements to Richardson that he would make good this time were not merely empty or hypocritical. He noted that addiction was a burden, that it decreased his energy and his capacity to enjoy life. It is easy to see how such a person would provoke reformers to adopt a more pessimistic stance toward addicts and why he would be an attractive target for law enforcement.

Michael's pattern of use reflects the pharmacology of both acute use of opiates and patterns of addiction. An individual consuming morphine or heroin on a steady basis begins to develop tolerance to the drug; that is, escalating doses are required to produce the desired effects. As the customary dose rises, however, the various costs associated with drug use also escalate. These include the increased cost of buying more of the drug, worse side effects, and diminished pleasure. So the addict often undertakes dose reduction or complete detoxification to get the situation back within more manageable limits, and this kind of habit management forms a frequent motive for seeking treatment. In the 1920s, "cure" was thought of simply as achieving abstinence, and since this could be accomplished by any regimen ensuring that the patient took no opiates for seven to ten days, the familiar revolving-door pattern of treatment was common. Achieving abstinence meant reducing tolerance back to levels associated with the period before drug use, so that a comparatively small dose of heroin or morphine could pack a big kick. Thus, any recourse to drug use provided a powerful experience of the euphoric and calming effects of the drug. However, tolerance quickly returned to earlier levels, and patients found themselves on an accelerating roller coaster of euphoric use, rising tolerance, diminished pleasurable effects, dose reduction, and, again, resumed use.

The Limits of Compassion

In contrast to Michael Heider, Patrick Collins represents the kind of case that frustrated physicians and officials struggling to implement the Harrison Act fairly and not deny pain-relieving medications to those who legitimately needed them. Collins's drug-using career spanned the decades from the first stirrings of Progressive Era concern over opiates to at least 1927, the year of his two admissions to Philadelphia General Hospital; from the time when a child could buy morphine in any drugstore and opium smoking was confined to the Chinatowns of large American cities to the time when use of opiates was a federal crime and a sick old widower with chronic back pain who had used opiates for thirty-nine years had to turn to the illicit market to supplement the meager doses a sympathetic physician provided for him.[8]

Patrick Collins was born in 1867 or 1868 (he couldn't remember which) to an Irish-born mother and her second husband. The family lived on Mott Street in New York City in a neighborhood whose Irish residents were being steadily displaced by Chinese immigrants. Collins's parents both drank, and, when drunk, they fought. He apparently received little attention of any kind from his parents; as he said, "My people did not have any control of me and they was of no account doctor. They drank too much. I did the best way I could and brought myself up." Asked about his first memory, he replied, "Just the drunks my father got on. My mother too." Patrick said of his mother, "She was not married long to any of her husbands." She and Patrick's father separated, and the father died of stomach cancer at Bellevue Hospital when Patrick was nine, leaving him with few memories of him. His mother was married for a third time, this time to a man "who did not have much use for me."

To get away from his unhappy home situation, Patrick developed an independent life as soon as he could, roaming a neighborhood that by his adolescence was being called Chinatown. He recalled fantan games and raucous Chinese New Year celebrations. At age 12, he began running errands for spending money, hiding his earnings from his parents; later, he sold newspapers at the entrance to the Brooklyn Bridge. As his parents exercised little supervision, he was soon taking most of his meals in restaurants and staying out late with friends. At 15, he moved out into a room of his own, "ran around the streets and played with the boys," gambling and shooting craps. Exposed both to ubiquitous bars, where he hawked his papers ("The corner stores would have a bar at the back"), and to opium dens, where he learned to cook up opium for money, the boy found he had little taste for liquor but that opium made him feel very good.

The occasion of Patrick's first opiate use, though, was a painful case of facial neuralgia in 1888, when he was about 20 years old. Someone suggested he take laudanum to help him sleep, and he did for several months. He liked the feelings that laudanum gave him enough that he continued taking the drug after the neuralgia abated. Then, when he began smoking opium, he found he preferred it to laudanum.

Patrick's neighborhood was, by his own account, a tough place, where a boy had to be sly to get by. He left it as soon as he was able, and thirty years later, he said that he had never had any interest in returning to it. In some ways, though, it was less tough than Richardson, undoubtedly drawing on his own sense of urban gangster life in the mid 1920s, believed. Richardson's questions probed for a Chinatown riven by violence and opium dens where orgies were common, but Collins drew a less lurid picture. Yes, there were fights, but he'd never seen anyone killed. Yes, women commonly took off their shoes and dresses in opium dens, with little concern for modesty, but they did so the better to relax as they lounged in the den's cubicles and smoked their pipes, not as a prelude to group sex—and they typically put on robes provided by the establishment.

At a certain point, Patrick escaped the family that had neglected him and the neighborhood where he had been forced into a rough premature independence. He followed some pals who had moved to Atlantic City to run an opium den of their own. Patrick found work painting houses and puttying windows. He moved on to other jobs: digging ditches for the Atlantic City Gas Company, carpentry, and washing bottles and cans for a dairy. The latter job lasted seventeen years, until the dairy owner fired Patrick so that he could give the job to his newly grown son. Patrick worked for several years at a factory that made hair ribbons, but the advent of bobbed hair cost him this job: "I quit when the girls got their hair cut, that way it put the business on the bum, there was no demand for ribbons." His most recent employment had been delivering ice. About two years before his admission to Philadelphia General, he had carried a load of ice on his back a long way on a cold, rainy day. When he got home, he sat by the window with his wet clothes on, and his back seized up with pain.

In about 1908 or 1909, he reported, he had switched from opium to morphine (although he did not mention it, this was about the time that federal law banned the importation of smoking opium). For the nine years preceding his admission to Philadelphia General, he had been able to get prescriptions of morphine from a physician, beginning with 90 grains a day. Patrick had apparently gained an exemption from the ban on medically managed morphine maintenance that was granted to old or chronically ill patients; but, he said, "the authority made him cut me down to 60 and then down to 30." This dose was not enough to keep Patrick comfortable, so he bought another 30 grains on the illicit market.

When he was about 35, Patrick married Margaret O'Bannon, a widow with three young children. She and her first husband had run a boarding house, where Patrick lived and befriended his landlords. Two years after Mr. O'Bannon's death, Patrick decided he was fond enough of Margaret to marry her. He believed he could be a help to her as she raised her children; he wanted to take care of her. Patrick's marriage was clearly his greatest source of happiness. The marriage lasted until Margaret's death a few months before he came to the hospital. Patrick often wept as he described their life together. He remained close to her two surviving children, living in the home of his grown stepdaughter, where he now helped her with her own children, taking them to movies and fixing meals for them while their parents both worked. His stepson provided him with partial financial support.

One aspect of Patrick and Margaret's pleasure in each other was their shared enjoyment of opiates. Like her husband, Margaret first began taking opiates for medical reasons. She had heart trouble and woke every morning with headaches. Her husband introduced her to a drug that relieved both pain and anxiety: morphine. She became dependent on it and continued to use it until the end of her life.

A social worker who visited the home where Patrick lived with his stepdaughter's family confirmed that she and her husband were happy to have the old man living with them. He was discreet about his drug use, confining it to his room, where their children would not be exposed to it, and his help in caring for the children made it easier for both parents to work at jobs outside the home. Officialdom was less sanguine about Patrick's habit, however, as evidenced in his doctor's repeated reduction of the doses he prescribed for Patrick. This physician may well have been afraid of prosecution under the Harrison Act, which was now being rigorously enforced, as Charles Terry and Mildred Pellens noted. Caught between the obligation to provide compassionate care for a sick old man and the stiffening rules about prescribing, Patrick's doctor may have felt compelled to reduce Patrick's dosage from the level that kept him comfortable. Physicians all over the country faced a similar dilemma in caring for aging patients who had become addicted to opiates before any law restricted their access to them.

Life Structure and Drug Use

Joseph Donatello's case allows a dynamic view of the progress of an addiction and how it interacts with other aspects of an individual's life.[9] This dynamism belies the static and essentialist nature of addiction implied by Richardson's assessments of his patients. Sociologists and urban ethnographers have developed conceptual tools for understanding addiction in the context of an individual's life course, identity, and social environment. A key feature of ethnographic methods is that they allow reconstruction of the addict's life in terms meaningful to the addict him- or herself. Howard Becker's "career of deviance" model, introduced in the early 1950s, was built upon by ethnographers in the 1970s and 1980s to produce a range of insights into the meanings of drug use.[10] By stripping away the judgmental lens of the physician, police officer, or moral entrepreneur, the career framework allows the story to be told in neutral terms. Eliciting the career creates a biographical structure along which to range successive episodes, events, and choices that add up to a coherent story. Attention to the social contexts of behavior suggests how choices are facilitated or constrained by structural factors such as law, social custom within a group, the nature of drug markets, and so forth. Framing a behavior pattern that deviates from mainstream norms as a career also permits comparison with more conventional kinds of careers; it highlights such issues as making a living and investing activities with a sense of identity. Such insights can be applied to the experiences reported by patients in Philadelphia General's narcotics ward, even given that the concerns that motivated Richardson's questions were very different from those that shape ethnographers' interviews.

Charles Faupel's ethnographic approach stresses two key aspects of addict life that are underplayed or ignored in Richardson's questioning:

its social nature (as opposed to Richardson's emphasis on individual behavior and character) and its dynamic character (as opposed to Richardson's attempt to fix the addiction in a static conception of psychopathology or personality type).[11] Faupel proposes a model that links fluctuations in heroin-use patterns to changes in the availability of drugs and to the amount of structure in an individual's life. Thus, when life routines are highly structured and drugs are not easily available, drug use wanes or ceases, as might be expected. Conversely, when drugs are extremely easy to get and a life is fluid rather than structured, drug use can easily spiral out of control. Two intermediate conditions are also possible. When life is highly structured around routines, responsibilities, and social roles, drug use may remain at controlled levels even given free, or at least regular, access to drugs. Finally, even when life structure does not provide supports for modulating drug use, if drugs are hard to get, scarcity may enforce moderate or sporadic levels of use.

Joseph Donatello was a first-generation American son of Italian parents who, except for a four-year stint in the Marines, had lived his thirty-four years in a working-class neighborhood in Philadelphia about twenty blocks south of City Hall. As a young child, he attended kindergarten and ran barefoot through the streets. One of his early memories was the curiosity sparked by the unusual look of some Chinese neighbors. As he moved through grammar school, he occasionally "bagged school" in order to play ball or swim. He completed ninth grade at age 14 or 15 and then quit school to begin working, first as a store clerk and then at a printing shop. At 18, Joseph was living at home with his parents, but he had considerable freedom of movement. At about this time, 1910 or 1911, he began to frequent Chinatown, some 25 blocks north of where he lived. This must have been on the eve of Mayor Blankenburg's reforms, because, in contrast to "today," he recalled a "wide open" Tenderloin where a couple of women he had known as girls in his neighborhood now lived and worked as prostitutes. Visiting them, he was offered an opium pipe to smoke. The motive for his first use was a desire to fit in: "Wanted to be a big fellow like the rest of them." He began a pattern of spending 24 or 36 hours at opium dens, smoking, lapsing into a stupor, and then sniffing cocaine to wake up. He claimed that within a week, he had developed a habit.

For Joseph, drug initiation passed quickly into steady use. He was introduced to the drug in a social setting by friends experienced with it. He was working a job during the day and living at home with his parents, apparently free not only to go out in the evening but to stay out all night.

His life structure thus offered no barriers to becoming a regular drug user: he had few responsibilities to anyone and considerable freedom in spending his time and money; and he had a home base, maintained by his parents, to fall back on after his opium sprees.

For a while, all was well. However, several factors altered this initial pattern. For one thing, the experience of the drug changed after he became a habitual user: "You only get fun out of it until you get a habit and after you get the habit you have to have it to feel good." In addition, the outside world intervened: he lost his job as a result of his habit. Then, about a year and a half after he began smoking opium, his customary "hop joint" (opium den) was raided, and he was arrested and jailed for eighteen days. He broke his habit (went through withdrawal) in jail.

The crisis of Joseph's arrest apparently precipitated a decision to alter his life dramatically. In about 1912, just after getting out of jail, he joined the Marines and served four years, part of it on a U.S. Navy battleship that stopped at ports in Asia, Africa, and Europe. Serving in the Marines created the opposite of his situation at home: life was highly structured and drugs were not easily available. He could have bought drugs in foreign ports, he indicated, but was afraid to do so. The combination of the discipline of the Marines and the lack of immediate social connections who used drugs created a situation in which Joseph remained abstinent for four years, the length of his service.

Joseph's return to Philadelphia after his discharge from the Marines created another reversal in life structure and drug availability: he had saved some money and decided to spend some time enjoying himself before looking for work; he resumed his social connections with his drug-using friends from before. By now, it was about 1916, seven years after the ban on opium importation and a year or so after the Harrison Act began to be implemented on March 1, 1915, and heroin had largely supplanted opium among his old crowd. One of his friends showed Joseph how to sniff heroin; Joseph said he hadn't known it was a habit-forming drug and was surprised that it made him feel much like opium had. In this completely unstructured period before he resumed work, Joseph used so much heroin that his tolerance quickly rose. Within three months, he doubled the amount he needed to buy each week, and after four months, he "started hitting the needle [because] I got more sensation from it."

After a time, Joseph got a job driving a delivery wagon for a fruit distributor at the docks and fell into a pattern in which he juggled the exigencies of his habit with his other responsibilities and roles. After two years, he quit this job to seek better-paying work. His father was a longtime road worker with the city highway department, and one of his brothers was a transcribing clerk in a City Hall office. Perhaps through one of these connections, Joseph also secured a job working for the city, punching cards. He had married (apparently while in the Marines), and in about 1920, a son was born to him and his wife. Over the years from about 1916, when he had returned to Philadelphia and begun using heroin, to 1926, when he was interviewed at Philadelphia General Hospital, his life formed a pattern of stints of work and drug parties interrupted by arrests, short jail terms, and treatment episodes in which he was detoxified and released.

For some periods, Joseph maintained a stable dose level and lived a regulated life: he took the same number of shots at the same times each day (delaying the first shot in the morning until he had had a bowel movement so as to minimize the serious constipation that can result from chronic opiate use) and kept his job. The routines of his drug use and the routines of his work life achieved an equilibrium, which included his need for a shot each morning in order to be able to work. But destabilizing factors continually disrupted these spells of regulated use. One such factor was undoubtedly a gradual upward drift of tolerance and dose, even in these periods of relative stability. Rising tolerance likely created a motive to enter treatment facilities, which he had done a number of times (including two previous stays in Philadelphia General Hospital). There were also forced interruptions; he was arrested often enough that the city jailers "welcomed me as a steady customer." While the first arrest had been a crisis, and going through unpalliated withdrawal on that occasion had produced real suffering, over time, he said, "I was getting used to breaking the habit . . . and did not mind it." Whenever he resumed use after an abstinent period, he started with a very low dose and got a big kick out it; and the tolerance fluctuation cycle would begin again.

Joseph's family relationships were strongly affected by his drug use. In the brief spells of abstinence, he would live with his wife and son and contribute his wages, some $28 a week, to the household. When he resumed drug use, however, he would move out and return to the home of his parents, citing his wife's unhappiness with his drug use and his own shame. His birth family also disapproved of his drug use, but for whatever reason, he felt better able to tolerate their responses than his wife's. Rising drug doses eroded his capacity to support his family; just before his current hospital stay, he had been spending about a fifth of his wages on heroin.[12] His wife worked in a cigar factory, and she was thrown back on her own resources to a greater extent, in any number of fiscal, logistical, and emotional ways, when Joseph was absent.

"When he is off the drug he is never contented. Always misses it," Joseph reported. Even when he felt able to enjoy life, he still missed the drug: "You're always craving something—always missing something, no matter where you would go or who you were with you feel lost all the time. Something pleasant missing." After about three months, though, a volatile state set in: "When he feels good he is a wild man. Wants to go this way and that and that is when he gets started on the stuff again." At a different point in the interview, Joseph indicated that he had not gone more than three months in an abstinent state since leaving the Marines. The point when mild depression gave way to a somewhat manic energy clearly represented a point of vulnerability for Joseph. Given an easy recourse to a social scene where heroin was used and accepted, and the lack of any support for continued abstinence that did not invoke guilt and shame, Joseph repeatedly resumed heroin use and began the cycle over again.

Joseph's life and addiction illustrate Faupel's model of the relationship of life structure and drug availability to patterns of use. His adolescent life leant itself to easy involvement with opium and cocaine. The four years he spent in the highly structured environment of the Marines, despite the extensive travel involved, were the only sustained abstinent period of his adult life. The spree following his discharge corresponds to what Faupel calls a freewheeling pattern: with a sudden shift to ample supplies of drugs (or money to buy them) and the lack of even a job to structure his days, he moved quickly from his first sniff of heroin to injecting high doses of the drug. A finer-grained account of the years following 1916 might indicate that his decisions to enter treatment and his moments of relapse to renewed drug use were modulated by events in his daily life or his relationships with his wife, son, or birth family.

Gender and Class

Richardson's interactions with Frances Simmons reveal his exhortative style of treatment, while her account illustrates how addiction interacted with other aspects of her life, including her marriage.[13] Frances first used heroin at a dance hall in 1915, when she was 19 years old. She complained of a headache to a friend of hers, a woman who worked at the dance hall accepting tips to dance with men. In the bathroom, the friend suggested Frances try heroin for her headache. When sniffing it provided no relief, the friend showed Frances how to inject it under the skin. Frances quickly became addicted and used heroin, and occasionally morphine, for twelve years, until 1927, when she came from her home in New York to the narcotics ward at Philadelphia General. She was feeling bad, she said, and had been losing weight for some time, so she wanted to get off the drug.

Frances had twice run away from home as a young girl. Later, she worked as a waitress and married James Simmons. Her respect for her husband did not prevent them from engaging in frequent conflicts. At one point, frustrated because she felt he did not spend enough money on her, she resolved to leave him. She and another waitress decided to audition for work on the stage. Frances said she hoped in this way to earn a larger independent income; however, the theater manager did not hire her, and she returned to her husband and her work as a waitress.

James Simmons resented the cost of his wife's drug habit. At first, she had saved enough money to buy her heroin from the money he gave her for groceries, but in time this did not cover her needs. Despite their arguments about money, she understood his point of view: "He has to work hard for his money and I don't blame him." When she decided to travel with a woman friend, also an addict, to Philadelphia to seek a cure, she did not inform her husband of her plans, but simply left him a note telling him she had gone away.

Richardson questioned Frances on this point: "Does [your husband] know where you are?" "Do you think he should know?" he asked. Dissatisfied with her answer, Richardson told her that she had done wrong in failing to inform her husband of her intentions and whereabouts. Later in the interview, he elicited the fact that Frances frequently went to dance halls without her husband, as she liked to dance and he did not. Frances did not believe this threatened her marriage: "As long as I do not do anything else but dance he says go dance all you want. He is not mean like that. If a man cannot trust you are not worthy to be trusted." "Are you sure your husband trusts you?" Richardson pressed. "Yes sir, I am positive," Frances replied. But Richardson saw peril in Frances's association with a woman friend she went out with in the evenings: "Josie had men in her apartment that were not decent didn't she." Frances acknowledged that her friend "did not do right" but refused to carry the conversation further.

When Richardson returned on another day for further interviewing, Frances complained, "Talking, my God, you talk to me yesterday until you had me hysterical. I am completely tired out." Richardson interpreted Frances's behavior as willfully obstinate ("You are always looking for a fight aren't you?") and told her, "I have dealt with girls like you for a long time." The stenographic record adds Richardson's suggestion that her cooperation with his research would be useful for others: "Examiner explains to her that he is only doing this to help people of her class." He then imparted his evaluation of her character and deployed his standard therapeutic method, a lecture on better living: "Examiner goes on to tell her that her sort seem to have one set of rules for themselves and one set for the outside world and tells her that her outlook on life is bad and to try and correct it." He also invoked her love for her husband: "How much do you really think of your husband?" "A lot," Frances replied. "You did not think enough of him to stay off the drug, did you?" He then urged her to reconcile with her husband and give up her other friends: "Examiner tells patient . . . that the people with whom she has associated heretofore have been worth less than nothing." He closed with a suggestion she attend her Episcopalian church more often. The inadequacy of his exhortations is evidenced by a letter Frances wrote the admitting physician on the ward twenty months later seeking readmission: "Just a few lines From one of your bad Patients Doctor. . . ."

Richardson's judgment of Frances reflected the common psychiatric view in the 1920s that equated mental health with conformity to conventional middle-class norms of gender, family, work, and social status. Frances violated these in a number of ways. Although she did not say she earned money at the dance halls she frequented, it seems likely that she did. She spoke of close friendships with women who did work in dance halls, and her attempt to get into show business may have reflected a desire to increase her income rather than simply to earn some spending money. In any case, she clearly desired more income and the autonomy it would entail, as the disputes with her husband over what he spent on her indicate. For women in Frances's position, an array of jobs that traded on female sexuality competed with such options as factory work. For a young, attractive woman who could socialize easily and dance well, work in a dance hall could mean substantial payment for a few hours' work several evenings a week.[14] Frances's husband's apparent willingness to let her go to dance halls without him may mean that he, too, saw this as income-generating work rather than fun. The costs of supporting a heroin habit can only have exacerbated the couple's tensions over money.

Frances's record lacks any note by Richardson indicating a psychiatric diagnosis, but his remarks to her and the stenographer's summaries indicate that he interpreted her behavior as typical of the kind of self-centered, short-sighted person he called a constitutional inferior. Going to dance halls without her husband, leaving him to seek work on the stage, and sneaking off to the hospital without telling him all violated Richardson's standards for wifely behavior. He saw her desire for some independent source of income as evidence of self-centeredness. His remarks about "people of her class" and his assessment of her friends as "worth less than nothing" betray class prejudice as well as concern over the risk of continued contact with addicts.

Everett Lewis and the patient referred to in the records as Dr. Y. exemplify middle-class addicts. Everett Lewis made several visits to Philadelphia General Hospital, the first in the summer of 1926.[15] He was an accountant by training and had a good job working as an accident claims investigator for an insurance company. When Everett was twenty-three, he observed a co-worker sniffing a powder and asked if he could have some. His reaction to this initial dose of heroin was to feel more active and ambitious than usual. Despite experiencing nausea the next few times he sniffed heroin, he continued to use it, and his friend continued to supply him at no cost. He increased his doses daily, and at the end of two weeks, he was sniffing four to five grains a day. One morning, he woke up feeling ill and ambitionless. Not knowing what was wrong, he mentioned it to "some of the fellows and they said I had to have it and it required more to make me feel normal." Over the course of about ten more weeks, his dose rose to ten to fifteen grains a day. He then entered Bellevue Hospital and took the belladonna cure (a variant of the hyoscine cure). He checked out after eight days (that is, at the time withdrawal symptoms had faded away) and moved to the country "to recuperate." He lived in upstate New York for ten years without using opiates. As in Joseph Donatello's case, a shift of scene away from drug connections created a context in which Everett was able to stay off drugs for an extended time.

When the plant where he worked closed, Everett returned to New York City. There, he soon ran into an acquaintance from his drug-using days and asked him for a shot. He then bought himself a supply and became addicted again. This episode started a pattern of periods of use interrupted by attempts to get a cure. Everett took the drug to help him deal with difficult or uncomfortable feelings. One relapse occurred when his first wife left him. He tried another cure but continued to feel despondent over his separation and resumed drug use.

Following separation from his first and second wives, Everett (again, like Joseph Donatello) lived with his parents, who became aware of his drug habit and strove to help him get over it. His father repeatedly made arrangements for Everett to enter treatment facilities. His mother became distraught each time she recognized the signs of Everett's resumed drug use, and Everett once started to cry during an interview when he was describing his mother's reaction to such an episode. At one point, Everett's parents drilled a peephole that enabled them to spy on their son in the bathroom to see if he was injecting drugs.

Everett's attempt to manage his habit reflect the dilemma of a man who tried to relieve difficult feelings with drug use but also recognized the negative effects of his habit. Morphine or heroin did not prevent him from functioning adequately at his job. In part through his father's influence, his job was always waiting for him when he returned from treatment episodes. "I have noticed that it has not interfered with my work in any way," he said. "Not many people know that I use it. . . . I never lost a position through using drugs." When he took drugs, however, he found that he lacked incentive or ambition to improve his situation. Off the drugs, he would become depressed and so start to use them again; on drugs, he became nervous and irritable.

Dr. Y presents a contrast in several respects to Michael Heider, Patrick Collins, Joseph Donatello, and Everett Lewis. Although Richardson interviewed him at great length, he does not appear to have been a regular patient of the addiction ward.[16] Rather, he was treated in another part of the hospital, and the occasion of his hospitalization was convalescence from a stroke some months before, rather than his addiction. He consented to be interviewed, despite feelings of shame regarding the addiction, in hopes that participating in the research would help others. While other patients were named in the clinical records, this and other physician patients' identities were protected; he is referred to simply as Dr. Y.

Dr. Y. was born in 1876 in a midsize Pennsylvania city, the only child of a mine supervisor. He was pampered by his mother and paternal grandmother, who bought him many gifts and who forbade him to swim at the local swimming hole where the rough Irish boys swam—although Dr. Y's own mother was Irish. Although barred from swimming, Dr. Y. spent much time outdoors, hunting chipmunks, roaming the forest, and firing his guns. Looking back on his childhood, he said that while he was the social center of a group of neighborhood children, this position resulted from his having more toys than anyone else rather than from personal popularity. At age 9, he began suffering from migraine headaches, and he was also plagued throughout childhood by stomach pains. After attending a private high school, he enrolled in a liberal arts college in eastern Pennsylvania, planning to study to become a mining engineer like his father. The father, however, forbade the son to follow his profession. So Dr. Y took science courses and then obtained his M.D. from an Ivy League university in 1901. He completed an internship at a hospital in his home town, and worked for several more years in hospitals in the surrounding area, primarily treating eye problems—an interest sparked by the frequency of eye injuries among miners and a history of eye injuries in his family. After working for several years as assistant to his fiancée's father, also a physician, he became a company physician responsible for compensation cases for a coal company. He treated coal miners, sometimes rushing to the mines when explosions caused injuries and deaths. He worked hard, frequently bolting his dinner and dashing back to work. He rarely took vacations. Although he continued to get painful headaches, he worked through them. He began taking Bromo seltzer or codeine for the headaches; after some years, he moved from codeine to the more potent opiate morphine, which he took orally and, eventually, by hypodermic.

Dr. Y's health began to deteriorate seriously when he suffered the first in a series of heart attacks in 1924. He then had an apparent stroke in 1926, an episode that left him paralyzed on the left side for a week and bedridden for about nine months. He was still in this period of convalescence when Richardson interviewed him.

Dr. Y's social class was reflected both in his own story and in his treatment by Richardson. Although Richardson asked Dr. Y about early interest in girls and his close relationships with women, the physician was not asked about sporting houses, masturbation, or venereal disease, as most male patients on the narcotics ward were. His favored pastimes included tennis and bridge, although his description of his work habits and his admitted discomfort with social situations suggest he did not spend much time at these activities. Dr. Y described himself as a worrier who took on everyone's problems. As a member of his local draft board during World War I, he agonized over decisions to send young men to war or exempt them from service. When his father-in-law died, he came down with an attack during a fractious family discussion regarding the will, and he professed reluctance to discuss such issues with Dr. Richardson.

Dr. Y thought of morphine as a medication, one he relied on primarily for managing the pain connected with his headaches. He said he got no pleasure or thrill from the drug. He recognized that he had become an addict and sought to end his addiction. In spite of encouragement from his own colleagues not to feel ashamed of his addiction, Dr. Y shared the reformers' view that his habit was shameful. This shame was one among many serious worries that beset Dr. Y. The interviews suggest he was always concerned about his health and that he was currently undergoing a crisis over whether he would ever become well enough to resume his work.

The influence of class on the experience of addiction can be seen in the cases of Dr. Y and Everett Lewis. With their professional or white-collar skills, both worked for companies willing to hold their jobs for them while they took time off for medical treatment of their drug habits. Dr. Y's serious health problems and prolonged convalescence made it unnecessary to provide any other excuse for taking time off; but Everett Lewis kept his habit hidden from his employers, making up other health problems that required hospitalization. Michael Heider, by contrast, had a spotty employment history, as a consequence both of his drug habit and his own predilection. He would give up jobs to seek a "cure" for his habit; he also quit work when the $600 his mother left on her death gave him the means to enjoy himself for a few months.

As a physician, Dr. Y was able to obtain drugs easily; in fact, he indicated that other physicians, including his own father-in-law, treating him for his various health problems, provided him with "hypos" of morphine for pain. Everett Lewis was able to protect himself from the vagaries of the black market by buying heroin in quantity, a month's supply at a time. Compared to someone without steady funds, he was less likely to run out of drugs and go involuntarily into withdrawal; and he was less likely to risk arrest, because he engaged in many fewer buying transactions than someone supplying himself from day to day. "I have not had any difficulty in getting it because I buy it by the ounce and therefore I could always get good connections," Everett said. "Those that buy it in small quantities would have to depend on street connections and are constantly being pulled in. Mine was delivered to me at the house."

Everett's continued employability and lower risk of arrest helped make him less visible to authorities than addicts buying drugs on the street. He was relatively successful at managing his addiction, in part because of his ability to keep a well-paying job and to fall back repeatedly on the support of his family. He used drugs to manage intolerable feelings, but his repeated insistence that this time he would stop using for good sound more heartfelt than those of Michael Heider. He was able to hide his addiction from employers. He was arrested once, not as part of a street bust, but in a peculiarly middle-class way: his calling card was discovered on the person of another drug user he had bought needles from, and police then sought him out. He supported the illicit market by buying his drugs there, but escaped the most serious consequences associated with it. The ability of this kind of addict to keep his habit relatively invisible, while those more impoverished or more closely associated with drug trafficking ran greater risks of illness, arrest, and public identification as addicts, contributed to the skewed view that policy makers and researchers formed of opiate addiction.

The Diagnostic Eye

The classifying eye of the psychiatrist grouped these addicts into a "type"—the "constitutional inferior" that appears repeatedly in Richardson's diagnostic notes. But the case notes reveal each addict as an individual, often isolated, struggling to manage a problematic habit while maintaining orderly relationships in several (often incommensurate) social contexts, including family, work, and the world of drug use. In the family, the addict often represented a problem, although in some cases, family members (brothers, spouses) also had drug habits. Many kept family separate from the network of drug users, which may have been an important social world for the addict, or just a set of contacts necessary for purchasing drugs. The addict thus faced a complex task of identity management, frequently including the need to pass as a nonuser with family or employers. The burdens of the drug habit often influenced work: the cost was a big issue, one that eroded the possibility of living or supporting a spouse or family on typical working-class wages. Some individuals chose forms of work that were more consistent with habit management—for example, occasional or self-employed work that would make it possible to take time off for cures periodically. Some opted for crime, which provided funds for buying drugs and probably also linked the user to social networks where drugs were bought, sold, and used. Many of these criminal activities involved hustle, the skills necessary to manage keeping a steady supply in an illicit market while avoiding arrest and minimizing conflict with friends or relatives who disapproved of drug use. For those who struggled to maintain respectability, learning to navigate the underworld to buy drugs added to their shame. For others, it proved compatible with their own desire for a hustling kind of life.

Many addicts found themselves moving between two social worlds, one in which addiction had to be hidden and another in which drug use was central. Often, family and work represented one pole and the site where drugs were bought (and sometimes used) the other. An individual might have different levels of involvement in each, or seek a balance between them. Some of the men living with their mothers, siblings, or parents were clearly very comfortable using this situation as a home base, from which they moved easily into and out of partying scenes where drug use was accepted. They still felt guilt over their family's (especially mother's) disapproval or disappointment regarding their drug use. This was probably a stronger form of a similar tension felt by many young adults (including perhaps especially first-generation Americans) in this period as they moved between a household whose older generation reflected traditional old-country or religious values and the social world of the dance hall and poolroom. Some lived almost entirely in one world or the other. Some were deeply ashamed of their inability to stay off drugs and please their families; many were frustrated by the difficulties of sustaining the qualities of drug effect that they had liked in the early phases of their use—either the pleasures or the sense of normality that drugs conferred on them. The individual moved among these conflicting worlds, trying to satisfy the demands of each; for many, the needs associated with the drug increasingly took over, in a pattern that the ethnographer Marsha Rosenbaum has characterized as a progression of narrowing options,[17] and led to the often repeated attempts to get "cured."

Richardson denied these complexities, attaching a single diagnostic label to the patient—usually, "constitutional inferior," a formulation of the "psychopathic" diagnosis that betokened, not full-blown disease, but the maladjusted personality type that was seen as giving rise to drug use and addiction. In this way, addicts fell into the diagnostic category that, as Jack Pressman shows, psychiatrists had developed as a means of demonstrating their utility in an urban America seeking to absorb a diverse immigrant population, adjust to the demands of a consolidating industrial economy, and manage social difference in a time of rising nativism.[18] Like the psychiatrists at the Boston Psychopathic Hospital that Elizabeth Lunbeck analyzes, Richardson aligned himself entirely with the "family" or other respectable center, and he purposely elicited the feelings of guilt associated with these figures or settings as a therapeutic tool.[19] For example, he frequently invoked mothers or spouses for whose sake the patient should try to "give up the stuff." For Richardson, the kinds of advice he gave about right living, the exhortations to be a man, to do it right this time, or to think of the guilt-inducing figure, represented his best therapeutic resource. For the addict, the hospital ward and its staff, including the physician researchers, became either another site of shame and guilt, or another place to hustle, a resource to draw on for habit management. By going through treatment and achieving (usually fleeting) abstinence, addicts could bring their customary doses down to more manageable levels, at least for a while. Agreeing to be studied on the ward meant several days of continuous morphine administration before withdrawal was undertaken.

Richardson weighed his patients against what Elizabeth Lunbeck has called a "metric concept of the normal," a graduated scale of deviance and conformity that became a tool enabling psychiatrists to deploy their skills, not just in the asylum, but throughout American society. The diagnosis of "psychopath" or "constitutional inferior" was used to label so many patients who exhibited poor social adjustment, personal distress, or peculiar behavior that it lacked any clear definition.[20] Richardson, like other psychiatrists in the 1910s and 1920s, encountered patients from an urban working-class world that deviated sharply from his own norms of class ambition, money management, sex, and gender; he tended to cast them indiscriminately as constitutional inferiors. He had trouble, for example, distinguishing single young women who worked in factories or stores, danced in the evenings, and sometimes took their dancing partners home with them from women whose main source of income was money for sex. He saw more similarity than difference between men who struggled to maintain supplies of opiates so that they could function better as printers or housepainters and street hustlers or professional gamblers. For him, all were people who thought primarily about themselves, who always put their own narrow needs first, who failed to plan ahead for their own futures, who led dull, circumscribed lives, and who lacked the inner resources of character to give up a bad habit even when they claimed they wanted to. These views were unchallenged by Richardson's own apparent fascination with the demimonde to which his patients gave him vicarious entrée—a fascination hinted at by his affectation of street argot and questions about details of gambling, carnival life, and gangsters.

The pattern of readmissions physicians observed in so many patients deepened their despair of providing them with either lasting relief or a definitive cure of a disabling medical condition. Repeated confrontations with the same patients professing renewed resolve to "make good this time" made them suspicious of the behavior patterns they associated with addiction, and of addicts as human beings. They increasingly saw the addict as an individual who lacked the symptoms of full-blown mental illness, but whose drug use occurred in a web of deviant behaviors. When the Public Health Service psychiatrist Lawrence Kolb published a set of articles in the psychiatric literature arguing that this pattern of addiction was a manifestation of inherent personality defect, many physicians were prepared to accept this explanation.

Roy B. Richardson's patrons at the Bureau of Social Hygiene saw little of value in Richardson's work and in time cut off funding for the psychiatric research altogether. This negative appraisal reflected concerns not only about Richardson himself but also about the uncertain status of psychiatry in the 1920s and the unclear boundary separating psychiatry from psychology. In the first year of the Philadelphia committee's work, Charles Terry expressed dissatisfaction with Richardson, but he favored continuing the psychiatric interviews, arguing that the Committee on Drug Addictions should not overlook the possibility that psychiatry might have something of value to contribute to an understanding of addiction.[21]

The committee dispatched the noted Harvard psychiatrist and committee member Stanley Cobb to visit the project at Philadelphia General Hospital, and he came away no more sanguine about the psychiatric work; as the minutes of the committee's meeting on December 17, 1927, record, "in the report of Dr. Richardson [Cobb] saw little to commend." Cobb noted that Richardson's training had been in neurology rather than psychiatry and thought this fact might help explain Richardson's difficulty in coming up with useful explanatory ideas.[22] In the 1920s, use by some of the label "neuropsychiatrist" reflected the lingering fluidity of the boundary between the two specialties. Since the 1890s, some psychiatrists had looked to neurology to explain the etiology of psychiatric illness. In addition, neurologists had the kind of office-based practice that attracted middle-class patients with mild forms of psychological distress (the so-called worried well) at a time when most psychiatrists, working in asylums with psychotic patients, sought a broader patient base.[23] But training in the physiology of the nervous system and diseases traceable to somatic nerve dysfunction did not necessarily equip one for the research task Richardson faced: determining what was pathological in the behavior of his subjects.

The ambiguous status of psychiatry in this period reinforced the Committee on Drug Addictions' doubts about the value of Richardson's work. A possible source of leadership that might have given a useful focus to the psychiatric work evaporated when the prominent psychiatrist Edward Strecker, citing the press of other duties, resigned from the Philadelphia committee that oversaw the research at Philadelphia General.[24] Charles Terry argued that both psychiatric and psychological methods "were as yet in their infancy and by no means on a scientific basis," and he came to favor cutting off funds for this aspect of the project at Philadelphia General. The committee agreed and stopped funding the psychiatric research at the end of 1927.[25]

The committee's support of psychological examination of addicted patients at Philadelphia General fared no better. Again, a mixture of personnel problems and disciplinary uncertainties were at work. The first psychologist hired left the project in midstream without submitting satisfactory reports of his work; his replacement also failed to meet the committee's expectations.[26] Cobb argued that the methods of psychology were not scientific.[27]

In one sense, the lack of consensus on a psychiatric or psychological direction left the field clear for other forms of research on addiction. The dissatisfaction with the psychiatric work at Philadelphia General occurred at the time when the Committee on Drug Addictions was opening negotiations with the National Research Council to take over scientific leadership of addiction research. When Reid Hunt's proposal to focus on the development of improved analgesics was put forward, no one on the committee was prepared to argue for a more psychologically oriented approach. Nevertheless, in another setting, the work of Lawrence Kolb, the 1920s saw the emergence of a psychiatric paradigm of opiate addiction that held sway for the remainder of the classic era of narcotic control.

5

The Junkie as Psychopath

Lawrence Kolb's formulation of addiction as a problem arising from the defective personality of the addict dominated medical, scientific, and policy thinking about addiction for several decades following the publication of his classic works in 1925. Kolb framed addiction in the terms of the new psychiatry, a reform ideology under whose banner psychiatrists sought to transform their specialty from an asylum-bound medical backwater to a foundation for commenting on and prescribing for almost every aspect of individual and social life.[1]

Kolb's theory that addiction resulted from underlying defects of character reflected the ideas of the new psychiatry being deployed in urban clinics as psychiatrists created new professional venues outside of the asylum for the insane.[2] His reputation rests both on the influence of his psychiatric explanation of addiction and on his prominence as an advocate for medical rather than criminal justice management of addiction. Ironically, Kolb, who spent the latter decades of his career opposing the imprisonment of addicts and arguing for medical treatment of their condition, also consolidated the picture of the addict as psychopath and served as first medical director of the Public Health Service Narcotic Hospital at Lexington, Kentucky, an institution that was both a prison and a hospital. The presence of these two strains in the same man illustrates the complex and transitional medical attitudes toward addicts in the 1920s. In that decade, a psychopathic model of addiction seemed an accurate description of the addicts who predominated in the waning days of city hospital narcotics wards like that at Philadelphia General Hospital. Faced with a pattern of continual relapse among those who sought "cures," physicians were losing confidence that addiction could be effectively treated. They developed a pronounced distaste for drug-hustling patients who complained of vague but intense pain that required treatment with morphine or codeine, or who repeatedly sought cures.

At the same time, physicians treating addicts in the 1920s were still acutely conscious of the distinction between those who had become addicted through medical treatment and those who, in their view, chose addiction by becoming involved in drugs as a source of pleasure. As the cases examined in Chapters 1 and 4 indicate, this distinction was not always as clear in the experience of addicts as it was in the minds of clinicians. But for clinicians, this divide was crucial. It marked off the patient deserving compassionate care from the vicious type for whom such care would do little good.

Lawrence Kolb strengthened this attitude by distinguishing "accidental" medical addicts from willful ones in psychiatric terms. For Kolb, those who became addicted through taking opiates to treat a medical condition not only lacked any taint of psychopathology; they also felt no pleasure when they took the drug. In contrast, those who took opiates for nonmedical reasons owed their subsequent addiction to defects of personality. As Kolb's assessment of individual cases will show, he was disposed to give individuals every possible break as he gauged their worthiness for exemptions from the ban on maintenance on the basis of chronic medical conditions that required treatment with morphine or that might make withdrawal dangerous. However, in his psychiatric writings, he rhetorically framed the addict who deserved compassionate care as a "normal" individual who had been caught up in addiction through no fault of his or her own, while he created a new psychiatric diagnosis to explain the behavior of an addict like Michael Heider (discussed in Chapter 4). This approach implicitly equated addiction with criminality for those who combined addiction with other forms of deviance. This construction of addicts as deviant was, in turn, applied to addicts whose only criminal activity consisted of possession or use of illicit drugs. Kolb's description of the psychopathic addict allowed this equation to be made by those interested in doing so, such as the federal officials charged with enforcing the Harrison Narcotic Act.

Larger currents in medicine and in American society also reinforced the distinction between the "innocent" medical addict and the psychopathic addict for physicians practicing in the 1920s. The extent to which medically induced addiction had already been displaced by the newer pattern was not clearly evident to many physicians, both those who commonly treated addicts and, especially, general practitioners. Many physicians still believed iatrogenic addiction was a serious problem. Hospital stays, including stays for painful traumas like serious fractures, were typically weeks long, and the risk of inducing addiction through the prescribing of morphine for pain in such a situation was significant. The American Medical Association was still deeply involved in a campaign to limit the prescribing of opiates and purge the profession of scrip doctors and physicians who were themselves addicted to opiates. Many, including Lawrence Kolb in the late 1920s, believed that effective enforcement of the Harrison Act would limit the pattern of deviant addiction to the cohort who had become addicted before or in the immediate aftermath of the implementation of the act, and that this disturbing type of patient would fade from the scene. But discouragement mounted as the number of addicts seeking treatment or getting caught up in the criminal justice system persisted, and recourse to institutionalization was an expectable response at a time when institutionalization or incarceration was being proposed and enacted for other kinds of stigmatized populations, such as the feebleminded.

What was dropped from this formulation was the psychopathic addict's right to the standard of humane care that is normally implicit in the doctor-patient relationship. Kolb's definition of addiction created a diagnosis that seemed to justify treating the addict as a criminal, and as federal policy toward addicts hardened under Harry Anslinger's direction of the Bureau of Narcotics, the criminality of the psychopathic addict increasingly overshadowed any conception of the addict as a patient deserving care. In this policy context, a diagnostic category that linked psychopathology so closely to criminality provided no rhetorical grounds for arguing that addicts should be treated with compassion and receive good care. Those who attempted to make that argument remained in a small minority until dissenting voices began to gain strength from the late 1950s on.

The fact that many addicts also gambled, frequented dance halls, and were sexually promiscuous or "deviant" reflected what Progressive Era reformers had discovered some twenty years earlier: various forms of vice typically coexisted, both in individuals and in neighborhoods. As a result of the social reforms enacted by Progressives and their successors in the 1920s, prostitution and opiate addiction were made illegal and driven underground. Red light districts were cleaned up, and the Harrison Act, with its anti-maintenance stance, was enforced.

The result was the separation of a tangled set of practices, and the spatial, social, and cultural contexts that supported them, into a different pattern, in which deviance was distinguished more sharply from respectability, even as the range of acceptable behavior broadened. Thus, by the 1920s, many Americans, including middle-class, respectable people who might have responded to the Progressives' calls to arms two decades earlier, accepted more liberal social and sexual mingling of young adults. Women wore shorter skirts and shorter hair without exciting disapproval. Dance styles that had once seemed lewd became popular. Music that had first gained an audience in brothels (ragtime from the late 1890s; jazz from the 1910s) became a staple of sophisticated urban entertainment venues in the increasingly complex harmonic forms developed by band leaders and composers like Duke Ellington, and in the simpler forms of boogie woogie and rural blues that enjoyed mainstream popularity in the 1920s. College students adopted more liberal social and sexual behavior after World War I.[3] John Burnham argues that cross-class mixing in the trenches of World War I exposed middle-class Americans to working-class mores and paved the way for the relaxing of behavioral standards in the 1920s; he shows the role of commercialization in changing the ways vices were purveyed and practiced.[4]

The repeal of alcohol prohibition in 1933, quickly followed by W. S. Van Dyke's enormously popular film *The Thin Man,* symbolized the legitimation of behavior that between the turn of the century and World War I had been seen as a sign of cultural dissolution. In contrast to the image of post-Repeal urbane sophistication portrayed in *The Thin Man* and its sequels, Dashiell Hammett's novel *The Thin Man* was set during Prohibition.[5] The inclusion of alcohol in virtually every scene creates a compelling portrait of Nick Charles as a late-stage alcoholic. However, the film *The Thin Man* portrays the bar as a place where chic urbanites deploy wit and display tony wardrobes. Nora's downing of five martinis in rapid succession and Nick's kidding response to her hangover the next day helped create an acceptance of heavy levels of social drinking.[6]

But opiate addicts and prostitutes were left on the far side of the newly drawn line dividing acceptable, if racy, behavior from that considered truly deviant. As a result, they were linked to crime and denied recourse to legitimate sources of help, and it was made harder for them to move from deviance into more conventional roles and identities. Progressive Era reforms had broken up the red light districts and scattered prostitution more diffusely through cities, although it still occurred primarily in working-class or deteriorating neighborhoods. The Harrison Act created fertile ground for the development of an illicit market, and addicts had to learn how to navigate that shady world to acquire opiates and manage their habits.

The Public Health Service and Psychiatrist Lawrence Kolb

Lawrence Kolb came to a career in psychiatry from a background in public health. A native of Galloway, Maryland, he graduated in medicine with honors from the University of Maryland in 1908, one of the last years in which a physician graduating from medical school typically had not attended college. The University of Maryland's medical school was an old-style proprietary school at a time when elite reformers in medicine were working to upgrade medical education and narrow the gateway to an overcrowded profession.[7] After a year as resident at University Hospital in Baltimore, Kolb joined the Public Health Service with the rank of assistant surgeon.[8]

Public health had been the first area to achieve significant practical successes following the discovery of the bacterial causes of such communicable diseases as tuberculosis and bubonic plague. Federal legislation in 1902 had renamed the Marine Hospital Service, which wielded federal quarantine authority as well as providing medical care for merchant seamen, as the Public Health and Marine Hospital Service. This change recognized the increasing power the federal government was assuming to control infectious disease. The Biologics Control Act, also of 1902, authorized the Public Health Service to regulate the production and distribution of vaccines and sera. Further legislation in 1912 would again rename the organization, which from this date became the Public Health Service (PHS), and authorize it to conduct investigations of disease, which it had in fact been doing at the Marine Hospital Service's Hygienic Laboratory since 1887. The Public Health Service bureaucracy was quasi-military in style, with a uniformed corps, lifelong career prospects, and a regular path for promotion.[9] Kolb was undoubtedly attracted by the combination of career security and work in an exciting area of medicine in which the recent revolutionary discoveries of bacterial pathogens and their modes of transmission were finding their most fruitful applications.

After an early posting at a quarantine station, Kolb was sent to Ellis Island, where he spent six years working chiefly in the mental examination of prospective immigrants. As patterns of immigration changed in the decades following 1890, the task of screening immigrants applying for entry into the United States became more complex. Barriers of language and cultural difference complicated the process of determining who was fit or unfit to enter the country. The development of intelligence testing and other psychological measurement methods provided means of setting standards and identifying individuals who appeared to lack the capacity to function as independent citizens.[10] The identification of those with often obscure indications of future mental problems was a challenge made to order for reform-minded psychiatrists seeking venues outside the asylum where they could deploy their diagnostic skills. In order to understand their preoccupations, and the kind of old-fashioned psychiatry that modernizing psychiatrists sought to distance themselves from, an examination of the nineteenth-century asylum alienist is necessary.

From the Asylum to the World Outside

Psychiatry in nineteenth-century America was the profession of asylum management and care of the institutionalized insane. In the United States, asylums, like hospitals, started out as institutions for those who lacked private or family resources in times of need. They were descendants of almshouses or charitable institutions for the dependent poor, with illness or insanity being one of the conditions that created such dependence.[11]

The form of the asylum crystallized in the United States in the mid nineteenth century, in part under the influence of Thomas Kirkbride.[12] He envisaged the asylum as a large institution organized like a community, where individuals who could neither function independently nor be cared for within families could live a well-regulated life that would help calm their disturbed thinking and behavior. Some would improve under such a regimen and return to life in their communities; others would pass the rest of their lives in the institution. Asylums on the Kirkbride model, including state hospitals for the insane, were built in the decades from between 1840 and 1880. They were typically located in the countryside, far from any particular city or town but relatively centrally located with respect to the geographic area they served.

Psychiatrists (called alienists in the nineteenth century) were charged with the care of the insane, but their functions did not typically include diagnosis or triage to the asylum. Patients were brought to the asylum either by distraught families who could no longer tolerate their aberrant behavior, or by means of committal through court procedures. Thus, the alienist's contact with the patient began when the illness was already well developed. In the asylum, the patient was housed in a hall or cottage with others of his or her sex and race; patients were also grouped according to levels of tractability or obstreperousness. Patients who were consistently violent or otherwise unmanageable remained in locked areas under the close supervision of attendants. Those able to function in a daily routine lived an ordered life of regular meals and activities deemed therapeutic.

Typically, these consisted of outdoor farm or craft work for men and domestic chores such as sewing or laundry for women. This regular and calming routine was believed to have a therapeutic effect that would promote recovery.

In such a scheme, in which medical care consisted of regulating the details of daily living, no clear boundary separated the administrative functions of institution management from therapeutic functions. Thus, the asylum superintendent involved himself in every detail of asylum life, from oversight of the medical care of patients to close management of the farm and household activities. Other physicians in the asylum held the rank of assistant superintendent and their upward career path consisted of rising to the position of superintendent as this became available in a given institution. Some women physicians were hired to care for female patients. The superintendent held virtually complete authority over institutional management, staff supervision, and patient care. Superintendents were eligible for membership in the Association of Medical Superintendents for American Institutions for the Insane (AMSAII). Its journal, the *American Journal of Insanity,* discussed issues of interest to the profession; these included not only advances in medical thinking about mental illness, but also the administrative issues involved in managing a large residential institution.

This system functioned well for several decades. In many cases these asylums offered, as Kirkbride had intended, a safe haven for troubled individuals, and a place where families could leave insane relatives and trust they would receive appropriate care. By the 1890s, however, asylum superintendents recognized that their profession faced significant problems. Asylums had not been able to demonstrate significant rates of cure. Increasingly, they were filled with patients who left only to return, or who simply never left. A survey of thirteen asylums in the early 1890s showed that 90 percent of their patients were chronic cases with no hope of release. The president of the AMSAII, John B. Chapin, lamented that "our asylums have become transformed into institutions for the chronic insane."[13] Better methods for identifying curables and separating them from the chronically ill were necessary. This would require closer examination of the earliest stages of mental illness; but psychiatrists lacked experience with such cases, because patients typically were not brought to asylums until their illness was full-blown.

Other factors pointed up the costs associated with the asylum's isolation from mainstream medicine and from major population centers. By the 1890s, the scientific revolution in medicine was well under way. Bacterial discoveries of disease etiology created new hopes for the control of disease, and as means of transmission of infectious diseases became known, state and local governments made substantial investments in public health activities. The discovery that many asylum inmates suffered from central nervous system complications of syphilis sparked optimism that other bacterial causes for mental disease might be found. But within psychiatry, little research of any kind was conducted, and this lack became a serious concern to leading alienists in the 1890s. They observed that in the related specialty of neurology, local lesions discovered at autopsy were being linked with forms of clinically recognized neurological malfunction. Had not the current wave of scientific reform begun with just such correlations between clinical signs and pathological anatomy in the early decades of the century?[14] Psychiatrists noted not only that neurologists were engaged in such research but also that their office-based medical practices brought them into contact with patients in the early stages of disease. Alienists' own asylums were filled with "clinical material," and the deaths of chronic patients provided ample cases for possible autopsy, but little advantage was being taken of these resources.[15]

In the early 1890s, a small group of reform-minded psychiatrists argued that better means must be developed for distinguishing acute, and perhaps curable, cases from chronic cases with poor prognoses. Calls for developing specialized hospitals just for such curables marked an early attempt to devise new institutional settings in which alienists could achieve contact with patients earlier in the course of their illness, before potentially curable cases developed into intractable ones.[16] In this decade, following the lead of European psychiatrists who were actively studying acute forms of mental illness, the AMSAII changed its name to the American Medico-Psychological Society.[17]

In 1894, the society invited the leading neurologist S. Weir Mitchell to address the plenary session of its annual meeting. In a powerful address, Weir charged that asylum psychiatrists remained mired in administrative minutiae, oblivious to the exciting currents of research that were transforming the rest of medicine, including neurology.[18] Some superintendents reacted defensively, arguing that their mission was chiefly the care and housing of large numbers of the indigent insane for whom society made no other provision. Executive ability was the chief qualification for such work, they maintained.[19] But others rose to the challenge, both acknowledging the need for more research in psychiatry and pointing to such research as was already under way within asylums.[20]

A group of young psychiatrists turned their attention away from the asylum and to the world outside. As the study of lesions failed to provide breakthroughs, they examined the work of Europeans like Emil Kraepelin who were adapting the testing methods of psychologists to the process of studying mental illness.[21] Again drawing on the work of Kraepelin and others in Germany, reformers urged the creation of urban psychiatric clinics (later called psychopathic hospitals) whose main functions would be preventive psychiatry and identification of emerging mental illness while cure was still relatively easy. The rural locations that had seemed essential to creating the desired bucolic atmosphere for asylums in the nineteenth century now meant that they were physically remote from the cities where preventive and early intervention work must occur.[22] In the two decades following 1900, the ferment of reform in medical education placed a premium on location close enough to academic centers to permit the creation of links between hospitals and university laboratories and classrooms. As reforming psychiatrists turned their focus to prevention and early intervention, other kinds of links had to be formed with surrounding communities: general practitioners who typically encountered the earliest manifestations of mental illness must be taught to refer cases to psychiatrists; school administrators and social workers must learn how to recognize cases appropriate for psychiatric intervention.

Adolf Meyer, a Swiss-born and European-trained psychiatrist who immigrated to the United States in the early 1890s, articulated a vision of the new psychiatry that adapted European nosological ideas and institutional innovations to the American scene.[23] His training had exposed him to Kraepelin's new, more precise and dynamic diagnostic categories, which incorporated the notion that a given mental illness might assume more than one behavioral or affective form, as in the case of manic depressive illness (bipolar disorder). Meyer also became familiar with the new psychopathic clinics that sought to identify cases of mental illness in their early stages and to provide early treatment that might preempt the need for institutionalization.

In adapting his ideas to what he encountered in the United States, Meyer developed a model that focused the aspirations of reform-minded psychiatrists while it addressed the American challenge of creating a national whole out of increasingly diverse social elements. Meyer's work and writings were extremely influential among psychiatrists; this influence reached beyond the profession in the activities of the National Committee on Mental Hygiene, a group founded by Clifford Beers, himself a former asylum patient, to promulgate new and less stigmatizing ideas about the treatment of mental illness.

Several components of Meyer's thinking provided a foundation for psychiatrists to expand their purview beyond the asylum walls and into virtually every social space. By the early 1900s, hopes that clinical-pathological correlations would reveal lesions crucial to understanding the causes of mental illness had begun to fade. In propounding the new concept of psychopathology, a European idea grounded in the search for early signs of impending mental disturbance, Meyer offered a concept of mental disease that did not depend on physical lesions for its reality. It was sufficient to note disturbed patterns of behavior or thinking through observation and mental testing.

Meyer also drew on the work of American pragmatist philosophers such as John Dewey and on the work of sociologists of the Chicago school such as Albion Small and George Vincent to develop a social model of mental health and mental illness. If psychopathology could be detected by observing an individual's behavior, the touchstone of mental health would be the individual's successful adjustment to the demands of his or her social role. In stressing the importance of adaptation to one's circumstances as key to determining personal success, Meyer was influenced by early functionalist models that saw society as comparable to a single organism consisting of mutually interacting parts, and that pitted social groups, including national or racial groups, against each other in a struggle for cultural survival. For Meyer and his fellow reformers, mental disorders, which could afflict anyone, became a focus of concern that made it possible to apply diagnostic skills to everyone in society. The focus on successful social adjustment not only provided a rhetorical standard for measuring mental health; it also laid the groundwork for psychiatry's professional expansion into myriad institutional settings, including schools, social service agencies, and prisons.

The U.S. Public Health Service became an important site for the implementation of the new psychiatry, especially through the work of Thomas Salmon (who served briefly on the Bureau of Social Hygiene's Committee on Drug Addiction). Salmon was a leading figure both in communicating Meyer's ideas to a wide audience and in developing institutional models for the deployment of these ideas. In 1904, Salmon was stationed at Ellis Island, where he took charge of developing a mental testing program to screen new immigrants so that those deemed mentally unfit could be denied admission to the United States. Through his work at Ellis Island and his publications based on that experience, Salmon developed a model of psychiatry as a public health activity. He argued that surveillance of the population and early detection of problem behaviors could have a preventive effect in reducing the incidence of serious mental illness.

For Meyer and Salmon, identifying those whose behavior betokened future problems thus became an important psychiatric function. Borderline individuals with a latent capacity for, or incipient, psychopathology represented a category of mental disorder, whose early treatment, or readjustment, meant a happier future for the individual and a significant savings in later treatment costs for society. Disorder, as distinct from full-blown disease, became both an organizing focus for ideas about psychopathology and a mandate for deployment of psychiatric clinics and psychiatric expertise in many social arenas. Psychopathy emerged as a broad, vaguely defined diagnosis, which was attached to individuals who psychiatrists observed were unable to adjust to the demands for self-restraint and social conformity called for by a complex society, but who lacked symptoms of severe mental illness such as delusions or profound depression.[24]

In 1913, Lawrence Kolb was stationed at Ellis Island, where he spent the next six years working primarily in the mental screening of prospective immigrants. He arrived at a time when Salmon and his team had come to recognize that differences of culture and language accounted for some of the behavior they had been calling feebleminded; they were working to improve mental testing to get a truer assessment of mental ability.[25] Ellis Island's proximity to New York City afforded opportunities for Kolb to get training at institutions where the new urban psychiatry was being developed and deployed. Kolb thus arrived in a setting where he was steeped in the new psychiatry. In the meantime, World War I intervened, and Thomas Salmon was called upon to head a team of psychiatrists that screened recruits and draftees and cared for soldiers deranged by their wartime experiences.

Psychiatric Screening in World War I

The massive mobilization effort required to assemble an army to fight in World War I and the problems soldiers encountered in combat created new roles and administrative structures for psychiatry. Psychiatrists were called on to screen recruits to eliminate those deemed unfit to serve. Asylum practice typically did not require distinguishing the sane from the insane. Furthermore, the screening effort placed a premium on identifying the superficially well adjusted individual who might, under the stress of combat, develop mental illness and then endanger other soldiers or remain on the public dole as a veteran injured while in service. The psychiatrists in charge of this screening effort, led by Thomas Salmon, developed training courses in diagnosing incipient or latent mental illness through examination of the behavior of potential or new recruits. Those who did not seem able to fit quickly into military routine or who stood out as eccentric or odd were included among the cases judged liable to later mental collapse.[26] Psychiatrists came out of the war with new experience in creating administrative structures and with new skills in searching large groups for individuals with signs of mental maladjustment.

World War I also provides an illuminating context for examining the response to addiction in an emergency setting where identifying addicts and removing them from the service was considered important for the national interest. The difficulty in recognizing addicts from among large groups of draftees and recruits posed a serious threat. The pragmatic need to develop a means of reliable diagnosis and the difficulties in doing so reveal something of why addiction became increasingly problematic at a time when efficiency, productivity, and cooperative social roles were highly valued by economic and social leaders. These Progressive Era virtues were especially crucial in the war setting. Addicts, however, were troublesome precisely because their condition could be hidden, yet disabled them from effective performance.

The need to examine tens of thousands of recruits and draftees for psychological fitness for service stimulated the mobilization of a large neuropsychiatric team and the development of fitness standards and diagnostic criteria. This effort occurred in the years just following the passage of the Harrison Act and before the Supreme Court had ruled that physicians could not prescribe opiates to addicts. Thus, it reflects ideas and practices prevalent at the time when federal legislation addressed the opiate addiction problem and before the government had moved to restrict physicians' freedom with respect to prescribing. Furthermore, it reveals a particular image of the addict that would reappear in the work of Lawrence Kolb: the addict as an individual who, like the psychopath, was chiefly noted for his inability to conform to the social standards medical officers subscribed to.

Volume 10 of the Surgeon General of the Army's massive *The Medical Department of the United States Army in the World War* (1929) surveys the effort to screen all recruits for mental fitness for service, outlines treatment resources and disposition of cases, and discusses the special psychoses and neuroses related to war (including shell shock). The need to identify prospective inductees with neuropsychiatric illness was critical because, if allowed to become soldiers, they would be liabilities. Even borderline cases, who might manage to get through life without medical attention under favorable circumstances, would certainly develop full-blown neuroses or psychoses in wartime conditions, when the lives of others and the national security depended on them.[27] Furthermore, it was important to identify such problem individuals before enlistment whenever possible, since discovery of a mental condition following enlistment meant that the soldier would qualify for government disability support. To allow such men to enlist rather than discovering their illness during preinduction screening would place an unnecessary economic burden on the state.[28]

The effort was hampered by the lack of psychiatrists experienced in diagnosing the mentally ill from among large groups of normal civilians, since the profession still consisted almost entirely of asylum superintendents whose careers had been spent working with the institutionalized insane. Medical education, it was noted, largely neglected psychiatric issues. Therefore, to ensure that there would be enough doctors to do neuropsychiatric screening, crash courses in psychiatric diagnosis were developed for physicians who had not been trained as psychiatrists and psychiatrists whose asylum positions had insulated them from diagnostic concerns.[29] Opiate addiction was a disqualification for military service.

Under a discussion of psychopathic personality, the authors spelled out the objective of the screening effort: "The question to be determined . . . at the examination of a recruit . . . is whether his make-up is such that his behavior, with practical certainty, will be inconsistent with service." Spotting the borderline case who might appear normal under some circumstances but who would crack under the pressures of military life and combat presented a special challenge. Careful observation of behavior was required to identify these individuals: "The behavior aspect is more likely to be noted than the health aspect, as is shown by the fact that soldiers are so often referred directly to psychiatrists by company commanders."[30]

Although every effort was made to identify the mentally ill before induction, further monitoring was required to identify those who had passed the initial screening and taken their place among soldiers awaiting shipment to Europe. In this context it was noted that, after induction, line officers were often quicker than medical officers to spot the behavioral abnormalities that betokened the personality disorganization that made a soldier unfit: "The explanation for this was found in the fact that the line officer rated his men in terms of conduct, behavior, and efficiency, which, after all, was equivalent to the standard of the neuropsychiatrist, who estimated conduct from the mental qualities and make-up of the individual."[31]

These types included those who could not fit into military routine and who thus would be readily apparent to the line officers: "Persistent delinquents, irresponsible, morally obtuse individuals. Eccentric, seclusive, taciturn individuals, company 'butts.'... Apathetic, negligent, untidy, or otherwise seemingly inferior or objectionable individuals."[32] Drug addicts were included in this list of types. The importance of fitting in smoothly was further emphasized: "Queerness, peculiarities, and idiosyncrasies, while not inconsistent with sanity, may be the beginnings or surface markings of mental disease. A soldier is too important a unit for such variations from a standard of absolute normality not to be looked into before the recruit who presents them is accepted for service." Thus, every officer must constantly be on the lookout for telltale signs among the enlisted men he came into contact with: "Irritability, seclusiveness, sulkiness, depression, shyness, timidity, overboisterousness, suspicion, sleeplessness, dullness, stupidity, personal uncleanliness, resentfulness to discipline, inability to be disciplined, sleepwalking, nocturnal incontinence of urine, and any of the various characteristics which gain for him who displays them the name of 'boob,' 'crank,' 'goat,' 'queer stick,' and the like."[33]

This laundry list of behavioral lapses reflects the diffuseness of the concept of psychopathy in this period. At the same time, the statement that the standard of the line officer was "equivalent to the standard of the neuropsychiatrist" was a potentially damning admission for a professional group striving for expanded social authority. That line officers could spot subtle signs of latent mental problems emphasizes the behavioral focus of diagnosis in a setting where absolute obedience and conformity were required. It also points up the flimsiness of the diagnostic categories psychiatrists were developing. Taken together, these factors indicate the focal importance of even minor forms of deviance as a problem that psychiatrists sought to exploit. The point that sticking out was a sign of not being able to fit in, although almost tautological, nevertheless captured what for the psychiatrists betokened incipient mental illness.

The World War I screeners described the drug addict in terms that combined these behavioral concerns with signs specific to opiate effects and opiate withdrawal:

> For drug addiction look for pallor and dryness of skin. If taking drug, the attitude is that of flippance and of mild exhilaration; if without it, it is cowardly and cringing. There are also, during period of withdrawal, restlessness, anxiety, and complaints of weakness, nausea, and pains in stomach, back, and legs. . . . Pupils contracted by morphine and dilated by co-

> caine. All habitual drug takers are liars. They do not drink, as a rule, and are inactive sexually. Most drug takers use needles and show white scars on thighs, arms, and trunk. Heroin takers are mostly young men from the cities, often gangsters. They have a characteristic vocabulary and will talk much more freely about their habit if the examiner in his inquiries uses such words as "deck," "quill," "package," "an eight," "blowers," "cokie" etc.[34]

This passage was clearly written by an experienced observer of addicts. These are almost certainly the words of the lead section author, Pearce Bailey, whose work with heroin addicts in New York City had made him familiar with the symptoms, behavior, and vocabulary noted in the passage (his observations on the shift from opium to heroin use were quoted in Chapter 1). Bailey's instructions suggest that he could spot heroin addicts quite easily. However, the terseness of his instructions and their setting in the broader context of troubling behavior suggest it would be difficult for a trainee psychiatrist (or nonpsychiatrist physician) to do so. These neophytes would probably have trouble throwing around terms like "quill" and "cokie" convincingly. They might well be forced to fall back on the importance of eccentric behavior, and in doing so they undoubtedly missed some addicts and suspected addiction in cases where it was not present: "By direct examination alone the only positive reliable evidence of habit are scars and abscesses from needle punctures. Failing to find these the most skillful physician can not be sure that addiction exists from any objective examination. In consequence, drug addiction belongs to the class of conduct disorders—the blight becomes evident more from the way the patient behaves than from medical examination."[35]

Another clear sign, but one that would appear only if the recruit were actively addicted and separated from his drug at the time of examination, was the appearance of withdrawal symptoms. Several days of observation would be necessary to clinch the diagnosis this way. Withdrawal was a defining aspect of addiction and of why it was an incapacitating condition:

> By a drug addict is understood one who has become so habituated to habit-forming drugs—chiefly derivatives of opium—that when suddenly deprived of them he falls ill with painful symptoms and can not work. This falling ill and inability to work is essential to the definition of drug addiction. Many, if not most, of the habitues, as long as they are supplied with what they have become dependent on, can work and keep in fairly good health. But they go to pieces shortly after withdrawal. It is in this way that

the diagnosis of drug addiction is chiefly made—not by direct examination, but by the so-called withdrawal symptoms.[36]

When the United States entered World War I, less than five years after the implementation of the Harrison Act, the addict was already being identified with "mostly tough young men from the cities, often gangsters." The linkage between addiction and the emerging concept of psychopathy created a syllogism for the screeners: social oddity constitutes incipient psychiatric illness; addiction manifests as social oddity; therefore, addiction reflects underlying psychiatric illness. When Lawrence Kolb moved from screening immigrants at Ellis Island to an intensive study of opiate addicts, he would elaborate an etiological theory of addiction based on similar logic.

The Home Front

Lawrence Kolb spent the war years conducting mental tests at Ellis Island. Like Thomas Salmon before him, he came to psychiatry from a background in public health and in a context that put a premium on identifying problem individuals among large population groups representing the whole potential range of mental health and pathology. While at Ellis Island, Kolb received psychiatric training, including six months at the New York Psychiatric Institute, a center of psychiatric reform and research. He also studied part-time for eighteen months at the Vanderbilt Neurological Clinic (also in New York). At Ellis Island, Kolb was engaged in the kind of public health psychiatry that characterized the movement led by figures like Adolf Meyer and furthered by Thomas Salmon. He was examining individuals representing the whole, nuanced range from superbly fit to seriously defective.

In April 1919, Kolb left Ellis Island to organize and manage a hospital for the treatment of nervous patients at Waukesha, Wisconsin. Then, in January 1923, he was assigned to the Hygienic Laboratory to conduct research on drug addiction.[37] He brought to this task his years of experience in screening population groups for evidence of mental unfitness and his exposure, through his training at the New York Psychiatric Institute, to the new ideas about psychopathology.

An early project was to estimate the prevalence of opiate addiction in the United States.[38] After compiling and analyzing such survey data, U.S. Army estimates, and such other figures as were available, and attempting to correct for errors, Kolb and his colleague Andrew DuMez, a Public

Health Service pharmacist, estimated that there had been about 100,000 opiate addicts in the United States in 1920.[39] Like most who had studied the problem of addiction in the United States, Kolb believed that the leading cause of addiction was excessive medical use of opiates.[40] He believed that passage of the Harrison Act in 1914, prohibiting sale of narcotics except as prescribed by a physician, adequately protected citizens from this hazard, and he predicted that the incidence of addiction would gradually decrease.

Meanwhile, he embarked on a study of about 200 addicts to attempt to learn more about the condition of addiction and those who were susceptible to it. He traveled to a number of Eastern cities and interviewed addicts in prisons and in hospitals; he also obtained the names of addicts not in institutions. In 1925, Kolb published his conclusions in a series of landmark articles in journals such as *Mental Hygiene*. He argued that the etiology of opiate addiction lay in psychoneurotic deficits predating any drug use by the addicted individual. While prior to 1914, many people had become addicted to morphine through careless prescribing, he claimed, the passage of the Harrison Act meant that now, only certain types of unstable or psychopathic individuals became addicts. Kolb characterized these as follows:

> The psychopath, the inebriate, the psychoneurotic, and the temperamental individuals who fall easy victims to narcotics have this in common: they are struggling with a sense of inadequacy, imagined or real, or with unconscious pathological strivings . . . and the open make-up that so many of them show is not a normal expression of men at ease with the world, but a mechanism of inferiors who are striving to appear like normal men.[41]

For these individuals, opiates provided a sense of well being that masked feelings of inferiority and allowed them to feel equal to their unrealistic ambitions.

Normal people who became addicted Kolb classified as "accidental addicts." Although, he believed, the Harrison Act effectively protected normal people from addiction by making opiates inaccessible to them, there was still the occasional case of an individual becoming addicted following prescription of an opiate for medical purposes. Two factors separated these normal people from the psychopathic addict: when they stopped taking the drug, they quickly recovered and did not relapse, as true addicts typically did; and they did not feel pleasure when opiates were administered.

Like other Progressive Era theorists about diseases linked to disapproved behavior in the early twentieth century, Kolb referred to "innocent" and "vicious" types to distinguish the two categories of addicts, the accidental, or normal, and the psychoneurotic, or psychopathic.[42] The boundary of addiction as a psychiatric disease fell between these two categories. Similarly, Kolb laid the disease boundary precisely along the cleavage created by the Harrison Act: whereas before the passage of the law, it was common for normal people to become addicted, he argued, since implementation of the law, only the "vicious" types did. Thus, barring the rare exception of the accidental addict, the opiate addict of the 1920s fell into a distinct diagnostic category. This psychoneurotic condition explained the etiology of addiction, and the addict himself was described entirely in terms of behavior and appearance: for example, Kolb compared addicts to "little men who endeavor to lift themselves into greatness by wearing 'loud' clothes or by otherwise making themselves conspicuous, when effacement would be more becoming."[43]

Kolb made astute observations regarding drug effects. For example, he consistently maintained that opiates, through their calming and sedative properties, were inherently more likely to discourage than encourage criminal activity; he also noted that cocaine "up to a certain point makes criminals more efficient as criminals. Beyond this point it brings on the state of fear or paranoia, during which the addict might murder a supposed pursuer."[44] However, these drug-specific factors did not enter into his discussion of the nature of addiction. His etiological explanations remained firmly planted in the framework of psychoneurosis and the psychopathic type.

Kolb described six categories of addicts. First, there were accidental addicts, normal people who had become addicted through medication during illness. The second group consisted of individuals with psychopathic diathesis. This group was the most ambiguously defined. Kolb believed the latent psychopathology of these individuals, under favorable conditions, might never become manifest. However, upon exposure to opiates, they experienced a profound relief of anxiety and a sense of pleasure that drove them to take the drug again. Once addicted, they had great difficulty in ending their drug use. If they did achieve abstinence, they were likely to relapse to drug use. The remaining categories were, third, individuals with ordinary neuroses; fourth, overtly psychopathic but not psychotic individuals; fifth, inebriates, or alcoholics, who became addicted when they took opiates to relieve hangovers; and sixth, individuals suffering some form of psychosis.

Kolb's approach was typical of the trend in psychiatry in the 1920s to develop classification systems that placed the source of problems within categories of psychopathology.[45] He developed criteria that excluded normal people and defined diagnostic categories consisting of defective individuals. This preoccupation with individual psychopathology as the source of social problems reflects a change from earlier critiques by Progressive reformers that emphasized the importance of environmental conditions in causing individual and social distress.

Many elements of Kolb's model were not new. Some thinkers had attributed the etiology of addiction to underlying psychiatric conditions since the late nineteenth century, and Kolb's portrait of the loud, garishly dressed "little men" echoes the descriptions of the soldier who couldn't fit in that appear in World War I neuropsychiatric screening accounts. Kolb used the term "diathesis," which had its roots in degeneracy theory, to refer to a latent predisposition for addiction that was revealed only when the individual began drug use and developed intractable dependency. However, in the climate of opinion of the mid 1920s, his ideas, formulated in the language of the new psychiatry, with its emphasis on social adjustment, reflected a new notion of innate defectiveness. While in the context of degeneracy theory, diathesis carried a fatalistic prognosis of progressive decline, framing diathesis in the context of possible readjustment to one's social role imparted a potentially optimistic note.

Implicit in psychiatry's acceptance of adjustment to prevailing social norms as a standard of mental health was an unquestioning acceptance of those norms. Certainly, psychiatrists worked to change norms in some areas; for example, an important component of the new psychiatry, in part as a result of the influence of Freudian thought, was an effort to overturn Victorian taboos on the discussion of sex and to allow free and open discussion of sexual topics. However, the aim was to help maladjusted individuals fit into prescribed social roles rather than to question any coercive or repressive aspects of those roles.

Kolb posited a psychopathology that was inherent in the individual and thus independent of the addictive drug, although contact with the drug was necessary to make the latent psychopathology overt. Therefore, a policy based on the goal of preventing entry of opiates into the country except as authorized for medical use could be viewed as a protective policy in psychiatric terms: it would enable borderline individuals to navigate their lives without encountering a hazard they were particularly susceptible to, and their chances of maintaining an appropriate adjustment were enhanced.

The contrast Kolb drew between the potential addict's overweening ambitions and his actual capabilities expressed a diagnostic view of addicts that accepted prevailing social norms. The implicit standard was one of individuals who would advance not through self-advertisement but through hard work at tasks suited to their narrow capabilities. They would remain content with their modest social status and adhere to middle-class notions of respectability while living in a working-class social context.

Kolb's views became the basis for constructing the addict as social deviant. His ideas were entirely consistent with the prevailing enforcement policy that called for arrest and imprisonment for possession of opiates. Kolb himself, however, did not endorse a punitive approach toward addicts. Rather, he supported the legal status quo because he felt the ban on possession kept dangerous drugs out of the hands of "normal" citizens. He also agreed with the position of federal enforcement and public health officials who opposed maintenance therapy for addicts, although he favored fairly liberal provisions for making opiates available to those with a medical need for them.[46]

Kolb consistently maintained that intractable addiction signified personality defects rather than any factor inherent in continuous drug administration. Individuals who took drugs for pleasure, "the pure dissipators," did so "because of unusual and pathological impulses . . . regardless of consequences and in spite of the penalties of laws."[47] This class of patients "relapse as a rule regardless of the treatment given," he maintained: "No elaborate system of rehabilitation will prevent this because the relapse is a mental relapse depending on their original pathological nature and due usually to the same causes that led to their original addiction."[48]

In contrast to this pessimistic assessment, Kolb did believe that treatment could help normal individuals who had become addicted through medical treatment. Records remain of Kolb's examination of thirty individuals from 1924 through 1928 while he worked at the Hygienic Laboratory.[49] These were people who came to him for examination to determine whether they qualified for exemption from the Harrison Act's prohibition on continued administration of opiates to addicts. In cases where a valid medical indication for long-term administration of opiates existed, addicted individuals could be classed as "legitimate addicts." Physicians prescribing for them would be specifically exempted from the Harrison Act's ban on continuous prescription of opiates for addicts; for these patients, maintenance would be judged a part of the physicians' "professional practice," the standard set by law for acceptable prescription of opiates.

These cases were not representative of the whole population of addicts; most were middle-aged and had begun taking opiates while under treatment for a medical condition involving pain or, less frequently, cough or diarrhea; approximately half of them had begun taking opiates before 1914. All had been referred to the Bureau of Narcotics for evaluation either by their physician or by a narcotic agent; thus, someone had already deemed them likely candidates for exemption. Twenty-four of the cases were male and six were female; six of the males were physicians. A number of the cases illustrate the ways in which typical medical practice of the period promoted the chances that medication with morphine would lead to addiction. Postoperative hospital stays of six weeks or more, with continuous administration of morphine for pain, were common. Patients were then sometimes sent home with supplies of morphine in tablet or injectable form to continue medicating their pain or other symptoms. When these supplies ran out, the patient discovered that he or she had become addicted.

Many patients managed to sustain relationships with physicians who would prescribe for them in the intervening years, especially if the physician had been trained before 1900. But as the Narcotic Division stepped up enforcement in the 1920s, physicians became more reluctant to continue this practice. Those who did prescribe for patients they had known and treated a long time became all the more crucial to these patients' continuing supply. Numerous factors could precipitate a crisis of availability; those cited by patients Kolb examined included the following: a patient's regular physician went on vacation, and the colleague who took on his cases during this period insisted on authorization from the Treasury Department to continue the patient's morphine prescriptions; either a physician or a patient moved to a new location; or a physician approached by a new patient presenting with a condition that might merit continuous morphine sought authorization before agreeing to prescribe the drug.

Many of these individuals had moved through cycles of attempted cure followed by relapse and resumed drug use for varying periods. The number of cures attempted ranged from a few to over a dozen. Such cures might involve stays in sanitaria for inebriates; they might be managed by the patient's physician, who would oversee gradual dose reduction; or patients might attempt such reduction on their own. For those who appeared before Kolb, all such attempts had ended in failure to sustain abstinence. Patients seeking legitimate addict status through examination by Kolb had to come to the Hygienic Laboratory in Washington, D.C., at their own expense.

Kolb examined these cases between 1924 and 1928, years that included the period when Roy B. Richardson was interviewing patients at Philadelphia General Hospital. The studies on which Kolb's influential articles were based had already been completed. He had articulated a clear distinction between normal, accidental addicts and vicious, thrill-seeking dissipators, a distinction that marked out, in theory at least, violators of the Harrison Act from legitimate medical patients. His job in examining these cases was to determine which of these statuses each individual represented. His disposition of them illustrates both Kolb's humane approach to individuals in this predicament and the social values that underlay his conceptions of normalcy. Nevertheless, his discussion of them contains clues that the distinction between "innocent" and "vicious" addicts was not as straightforward as implied in his published articles. Kolb's generosity in granting legitimate addict status and the rationales he marshaled in individual cases anticipate his more forthright rejection of criminal sanctions for addicts and advocacy of maintenance later in his career.

Kolb granted legitimate addict status to twenty-one of the thirty cases; in four other cases, he recommended temporary maintenance until the individual was able to stabilize financial or other concerns and then enter treatment for the addiction at a later time. Two cases Kolb rejected because he did not believe the individuals were addicted to opiates. In one case, discussed below (Dr. F.), the result was ambiguous; for another, Kolb's report is missing. Kolb recommended outright rejection of maintenance in only one of these thirty cases (Mr. B.). A close look at four cases illustrates their variety and Kolb's handling of them.

Mrs. L. represented for Kolb a clear-cut case of a legitimate medical addict. She was a 55-year-old widow who had first been given morphine at age 12 for headaches. As she suffered frequently from migraine, she had continued taking morphine by mouth, although not continuously. Then, following an operation for an ovarian abscess in 1900, she had taken the drug regularly and had become addicted. Her father had been a heavy drinker; her maternal grandmother, an aunt on her mother's side, and her brother had all suffered from migraine. She had married at 16 and had three children. Her domestic life had been happy; her husband had died in 1920. She had attempted cure three times, but each try had failed because as her drug dose went down, the severity of her headaches increased. She had begun to have trouble obtaining supplies of morphine eighteen months earlier when she had moved to another town.[50]

Kolb pronounced Mrs. L. a "very legitimate" addict on the following grounds: she had become addicted before passage of the Harrison Act; she had a clear medical condition, painful migraines, that merited treatment with morphine, the most powerful known analgesic; she had attempted cure numerous times but had failed each time to sustain abstinence; and she had none of the characteristics that suggested vicious use (for example, she had never purchased drugs from a peddler on the illicit market). He noted that Mrs. L. "has always been a moral woman and a useful citizen." The recurrent migraines made treatment of her addiction inadvisable, since morphine would continue to be indicated for the pain. Nothing would be gained by a cure, because relapse was inevitable. Kolb expressed no concern about the one point in her history that might have suggested that Mrs. L had a susceptibility for addiction: her father's heavy drinking, which Kolb might have taken as evidence of inebriety in her family history. This issue, if it troubled Kolb at all, did not outweigh the countervailing factors, which included a much stronger family history of migraines.

Other cases were more ambiguous but still merited exemption in Kolb's view. In a number of cases, Kolb's observation that the individual had replaced drinking with morphine use and functioned better as a result entered into his decision to permit continued morphine. Mr. E., a 60-year-old white male who had been a farmer and now ran a small store and filling station, had become addicted to morphine seventeen years earlier when a physician prescribed it after he injured his back in a fall from a tree. He had later received prescriptions to treat his rheumatism, and then had begun buying the drug himself. He had been cured three times, once staying off the drug for six months; but each time he had resumed taking the drug to relieve feelings of heaviness, lack of energy, and rheumatic pain. His father and one uncle had drunk heavily. He had begun drinking at age 11 and had gone on drinking sprees as a teenager; his history included some fifteen arrests for drunkenness and disorderly conduct. He had not drunk at all since beginning his use of morphine, except during the six-month interval between periods of use. He had married at 22 and had had three children. Following the death of his first wife, he had remarried. His second wife wished him to continue taking morphine, because, in her view, he got into less trouble since he had begun taking the drug. He had been a hard worker all his life and had bought a farm with his own savings. He reported that morphine brought back a boyhood capacity to enjoy life.[51]

Kolb reported that Mr. E. was lame but otherwise healthy and mentally normal. In spite of his drinking, he had worked hard and made a living. Permanent cure of his addiction appeared impossible because of the recurring rheumatism and because of his personality, which included a strong inebriate impulse; thus, Kolb cited both the legitimizing medical condition, rheumatism, and the propensity to addiction indicated by the drinking in supporting the case for continued morphine. Kolb employed a lesser-of-two-evils approach, saying, "By keeping him away from drink, morphine has . . . made a more stable citizen of him." Kolb pronounced Mr. E. a legitimate addict.

The case of Doctor F., a physician, reveals Kolb's assumption that a respectable professional and personal history would be incompatible with the kind of intractable addiction he attributed to "vicious" addicts. Kolb examined Dr. F. on January 26, 1926, and concluded that he was an entirely normal man who would not take opiates for purposes of dissipation. He recorded the following history: Dr. F. had become addicted in 1913 through medication of pain arising from chronic appendicitis (whether the appendix was removed is not stated). He had taken a cure in 1914, relapsed in three months, and taken another cure in 1915. By 1918, he was both taking morphine and drinking an average of six glasses of beer a day—an amount that did not seem excessive either to the doctor or, apparently, to Kolb, since his report included the statement that Dr. F. "has not been a drinker." Then, for several years, Dr. F. maintained a successful medical practice, keeping both his morphine and his alcohol intake at moderate levels. It appears that he had sustained abstinence for a period leading up to 1925, when he borrowed a significant sum to upgrade his medical practice and move it to an affluent neighborhood. Financial strain began to take a toll, and during a difficult obstetrical case, which lasted three days, during which he did not even change his clothes, he started taking morphine again. His morphine addiction was not interfering with his medical practice. He wished to be cured but did not want to interrupt his work during the lucrative winter season; rather, he sought authorization to continue taking morphine until the summer, when he could interrupt his practice with less loss of income.[52]

Kolb believed Dr. F. could be cured easily; he agreed that it would be unwise to attempt cure in such a way as to disrupt the doctor's work, especially as financial worries had contributed to his recent relapse. Follow-up correspondence, however, indicated that Dr. F.'s problems were not over. In June 1927, the physician's wife and mother telephoned Kolb in alarm, and Kolb paid a visit to Dr. F. He learned that following the planned cure, Dr. F. had contracted pneumonia and resumed use of morphine. He had then begun taking cocaine in the belief that it would make withdrawal from morphine easier. He reported that another physician had diagnosed a cancerous growth on his tongue, for which this doctor had prescribed an irritating topical ointment. Dr. F. cited the resulting mouth pain as a reason why he continued to take cocaine, applying it to his gums for its local anesthetic effect. This method of application would also produce cocaine's psychoactive effect as the water-soluble drug diffused through the oral mucosa and into the bloodstream. Kolb's final note on the case, while it does not reveal its disposition, indicates that he doubted the cancer diagnosis and now felt permanent cure of the addiction was unlikely. He urged the physician to give up using cocaine, whatever he did about the morphine, but Dr. F. insisted on continuing the cocaine at least for a few days, as he said, to control the mouth pain caused by the topical ointment.

Even at the first examination, some aspects of Dr. F.'s case might have curbed Kolb's optimism. He did not appear to be disturbed by Dr. F.'s pattern of drinking, which included periods of drinking six beers a day. While his theoretical distinction between medical and vicious addicts insisted that "normal" individuals felt no pleasure from morphine, he did not react to Dr. F.'s statement that in the early phases of morphine use, he felt exhilaration when he took the drug; this pleasurable feeling faded when addiction took hold. When he relapsed after his early treatment episodes, Dr. F. reported, "a hypo would cause blood to tingle through whole body, tingle all over, feel equal to any task." Coming from an unskilled laborer or streetwise tough, this language would undoubtedly have invoked for Kolb the image of "little men who endeavor to lift themselves into greatness." Meanwhile, Kolb readily believed that Dr. F. was capable of performing the challenging and delicate tasks involved in medical practice while maintaining a stable dosage of morphine; his recommendation that Dr. F. be allowed to continue taking morphine while he worked through the busy season of the year and postpone treatment until the slack season attest to this belief. The idea that opiate use at a stable dosage was compatible with maintaining a job might have suggested the usefulness of maintenance as an option for a broader range of addicts who were unable to sustain abstinence, but Kolb did not make this extrapolation.

In the end, the case bore the hallmarks of an entrenched pattern of addiction. Worried family members were seeking help for the physician. He, in turn, had added cocaine to his drug use and was telling a dubious story to overrule Kolb's advice that he at least stop the cocaine use. Dr. F.'s story appears dubious, not only because Kolb failed to find convincing evidence of the cancer, but also because Dr. F. had shifted from one justification for cocaine use—to ease morphine withdrawal—to another—to relieve mouth pain. Moreover, as a physician he was undoubtedly aware that by the mid 1920s, cocaine had been largely supplanted as a local anesthetic by other drugs that lack cocaine's euphorogenic property. Despite his respectable professional status and his ability to maintain a medical practice for a number of years while managing an intermittent addiction to morphine, Dr. F. became as intractably addicted as some of the street toughs Kolb had classified as vicious addicts. He had resorted to standard stratagems to hide his addiction: he had resumed morphine use during the pneumonia episode without informing his wife and mother (one can only infer the process whereby they discovered his lapse, but the example of Everett Lewis's parents drilling the hole in the bathroom wall to observe their son is typical of the kind of surveillance that distraught family members adopt when they suspect a relative's relapse to the use of drugs); and he had provided inconsistent rationales for his cocaine use. That Kolb handwrote "Not included in series" at the top of his initial report on Dr. F. may reflect his own confusion about this case. Yet Kolb's advice that Dr. F. stop the cocaine use regardless of what he did about morphine indicates his willingness to give pragmatic advice based on an incremental approach to improvement: even if Dr. F. was not capable of ending his morphine use, he should stop use of cocaine, the drug more directly associated with vicious patterns of use and more disorganizing in its behavioral impact.

Mr. B., the only case of the thirty that Kolb rejected as being ineligible for maintenance, although addiction was clearly present, did not present with the clear markers of respectability that Dr. F. had, despite some similarities in their histories. Mr. B. was a white male, aged 39, whose history included heavy drinking, gambling, and opiate addiction. Between 1913 and 1917, he had progressed from alcohol to paregoric (which contained alcohol and opium) to morphine, and his addiction to morphine, interrupted by six failed attempts at cure, had lasted until the present time. Drinking and gambling had caused him to lose money and jobs. At one point he was arrested for embezzling from an employer; his sister repaid the misappropriated funds.

Mr. B. claimed to derive no pleasure from morphine, and he attributed his relapses to pain and melancholy. He had worked fairly steadily as an adult, despite the jobs lost through drinking. He was married and had four living children. In the preceding three years, he had failed to support his family adequately because of problems in obtaining drugs; for example, he periodically left on trips to seek drug supplies in other localities. Mr. B. manifested no physical condition requiring medication with morphine. He was in early stages of withdrawal when Kolb examined him.

He requested authorization for six months' worth of morphine so that he could work and provide some money for his family; then, he said, he wished to enter a hospital for a six-month stay to attempt cure.[53]

Kolb judged Mr. B. physically healthy and intellectually normal. He noted that the history of drinking indicated an inebriate impulse that made relapse likely, but he agreed to support Mr. B.'s temporary maintenance while the latter organized his affairs. Of the thirty cases, this one received Kolb's harshest judgment: he believed cure was possible but unlikely, and as there was no physical illness to justify long-term morphine, sustained drug administration should not be allowed.

This verdict fell squarely within Treasury Department guidelines regarding exemptions to the ban on maintenance. Kolb was willing to give people the best break he could, but he did so within the boundary set by the 1919 Supreme Court decisions on the Harrison Act. In the case of the farmer-turned-gas-station-owner, he argued that morphine would be useful in keeping Mr. E. from drinking; and Mr. E. also had a physical ailment, rheumatism, that qualified him for maintenance under the terms of the Treasury Department rules. Kolb cited Mr. B.'s history of drinking as evidence of an inebriate impulse that would likely undermine his attempts at cure, but absent a qualifying medical condition, he would not authorize legitimate addict status. Kolb made no comment on Mr. B.'s claim that he derived no pleasure from morphine use.

Finally, Kolb's claim that a six-month stay at a sanitarium offered the best chance to cure Mr. B. attests to his belief that prolonged institutionalization was necessary to break the pattern of addiction and achieve stable abstinence. Six months in a hospital could only be seriously disruptive to Mr. B.'s family life and his ability to provide for his wife and children, but he had the misfortune to be addicted at a time when lengthy institutionalization was coming to seem the only treatment likely to work. Kolb explicitly rejected the usefulness of having Mr. B. enter a Washington, D.C., hospital for a short stay (undoubtedly to manage the withdrawal period, much as was done on the narcotics ward at Philadelphia General Hospital). Again, Kolb's own judgment that Mr. B. was capable of working while addicted to morphine for the period before he entered treatment did not extend to a recommendation that repeated treatment failures might justify morphine maintenance for this individual. And Kolb apparently thought Mr. B. was not a full-blown psychopathic addict, because he believed a lengthy treatment stint stood a chance of curing him. Thus, Mr. B. was caught, in a sense, between the two possible extremes: he was not an "accidental" addict who deserved maintenance, but he also, in Kolb's view, was not so lost to character defect that he might not benefit from lengthy institutional treatment.

In only one case did Kolb make a recommendation that skirted the boundaries laid down by the Harrison Act: he recommended maintenance for a man who became addicted to morphine through "having it prescribed for sprees."[54] Kolb attributed the man's binge drinking to a "constitutional psychic defect," which also accounted for the failure of four treatment episodes within a year to result in lasting abstinence from morphine. Kolb made the following argument to justify his recommendation that the man be allowed three grains of morphine a day:

> Even though this man has no physical disease requiring the use of morphine, the constitutional defect that impels him to take it regardless of consequences is a serious condition that should be taken into consideration. I believe that in his case the constitutional defect should be considered as creating a medical need for morphine, and that the interest of society as well as his own interests would be better served if he is allowed to have three grains a day so that he can attend to his business and be spared the demoralizing effect of dealing with peddlers.

This argument is striking in several respects. Kolb's theory of addiction held that a constitutional defect accounted for nonmedical morphine addiction, and he explicitly attributed both of this man's addictions—to alcohol and to morphine—to the same constitutional defect. In effect, he was allowing this man's addiction to justify maintenance. This case runs counter to the ironic trend in which addiction was defined as a disease requiring medical treatment and yet explicitly excluded as a condition justifying maintenance on the basis of medical necessity. The justification Kolb made for maintenance to a certain extent resembles the justification offered for methadone maintenance when it was introduced in the 1960s: keeping an intractably addicted individual maintained on stable doses of an opiate would help him or her avoid connections with the illicit market and make it easier to maintain social roles and obligations. Kolb's language also captures elements of a definition of addiction that would later overturn the insistence on character defect as an essential antecedent to addiction: he noted the compulsive nature of the man's drug use and drinking and indicated that these behaviors continued "regardless of consequences." Compulsiveness and use that continues in spite of adverse consequences are central to a definition of addiction that emerged in the

1970s and rejected the explanation based on a particular type of character defect.

This exceptional case notwithstanding, at this point in his career, incurability of opiate addiction alone did not justify maintenance in Kolb's mind. These cases reveal Kolb's beliefs about managing addiction when he had just consolidated his theory of psychopathic character defects as the etiology of nonmedical addiction. Kolb clearly did not intend that people should go to jail simply for being addicts; his support of the Harrison Act in the 1920s was based on the assumption that effective suppression of the illicit market, coupled with the ongoing reform of physicians' prescribing practices, would protect medical patients and those with a psychopathic predisposition to addiction alike by preventing addiction in the first place. Over the ensuing decades, as he witnessed the locking up of addicts who had become embroiled in the illicit market (as buyers and, often, as small-time sellers), he concluded that the Harrison Act had failed to eliminate the illicit market and had resulted instead in the futile and cruel incarceration of addicts for behavior they could not control. His examination of individual cases seeking exemption from the ban on maintenance occurred before such doubts had developed, but his discussion of them contains seeds of his later advocacy of maintenance. Regarding these cases, he stayed true to the Narcotic Division's insistence that only those with a medical condition apart from the addiction merited exemption. Nevertheless, as he examined individuals' lives and sought to determine the best and most humane disposition of their pleas for continuing morphine prescriptions, he voiced a pragmatic calculus that made an individual's capacity for social usefulness, rather than simply a psychiatric diagnosis, the deciding criterion in ambiguous cases.

On the one hand, Kolb was a key part of the policy-making structures that constructed the addict as criminal. His argument that addiction resulted from a psychopathic character disorder bolstered the view that addicts were inherently different from "normals"; thus, even if "cured" of a particular episode of addiction, they would retain a psychopathic taint. This view, coupled with his belief in the need for long-term institutional treatment to achieve cure in such cases, proved consistent with imprisonment of addicts.

On the other hand, Kolb's justifications for his verdicts regarding individuals' applications for legitimate addict status, along with some of his other writings of this period, contain the seeds of ideas that would become prominent decades later, as critiques of the management of addiction by the criminal justice system mounted. In 1956, the *Saturday Evening Post* published an article in which Kolb deplored the hysteria about addiction that had led to the passage of laws that, in his words, made a third drug-trafficking offense "the moral equivalent of murder and treason" (he referred to the Narcotic Control Act of 1956).[55] To counter the stereotype of the emaciated, hollow-eyed, criminal addict, he invoked a hypothetical respectable, if pitiable, woman of the 1890s who had become addicted to laudanum but who did not face criminal sanctions for her habit, and he argued that many addicts deserved the same pity such a woman did. He stressed that stable doses of opiates were entirely compatible with work. He deplored the pattern of enforcement in which addicts who peddled drugs to support their habit were locked up along with traffickers who were not addicted but profited from the misery of others. Finally, he argued that maintenance should be available to addicts while they awaited, or prepared for, treatment, and that incurable cases be allowed indefinite maintenance on stable doses of morphine. All of these points are present, at least implicitly, in his discussion of the cases he examined for legitimate addict status, save that he had earlier rejected maintenance for incurable addicts lacking a justifying medical condition, while now he saw incurable addiction by itself as grounds for allowing maintenance. He still attributed the intractability of the addiction to a neurotic failing of the individual, but he had ceased to believe that incarceration would be helpful in such cases.

In addition, Kolb anticipated ideas that became prominent in the decades following his death in 1972. In judging some drugs (alcohol, cocaine) more harmful than others (opiates), he rejected the dichotomous approach embedded in a legal regimen that banned certain drugs absolutely but tolerated even heavy use of others. Both the grassroots movement of the 1970s to provide straightforward information about psychoactive drugs to users and the more recent harm-reduction movement have argued the importance of distinguishing relative levels of risk connected not only with individual drugs but with distinct patterns of use of the same drug. Kolb's exhortation to Dr. F. to give up cocaine even if he continued using morphine and his belief that Mr. E. was better off taking morphine than drinking alcohol exemplify this pragmatic approach. Kolb's repeated assessment of social usefulness as a crucial criterion in determining whether maintenance would be appropriate anticipated more recent criteria for diagnosing drug dependence based on the relationship of drug use to other aspects of an individual's life (job, family, and so forth).

The Public Health Service Narcotics Hospitals

In position papers prepared for the Public Health Service, Kolb stated his continuing belief that incidence of addiction was not rising significantly, and that vigilance and careful public education would keep the problem under control. He argued that almost any treatment method succeeded in achieving abstinence, given two weeks or so in which it could be ensured that the patient would not receive any opiates.[56] For example, addicts in prison, cut off from drug supplies, began gaining weight in a couple of weeks and left prison in good health. That they immediately reverted to use of drugs did not reflect harm done by the drugs themselves, or any failure of the prison system, but indicated the presence of an inadequate personality type, whether psychopathic, neurotic, or inebriate. Therefore, when a bill came before Congress to authorize the construction of special prison hospitals for addicts, Kolb saw no reason for such institutions:

> If addicts possessing narcotics in violation of the law deserve a sentence of one or more years' confinement, there is no reason why the sentence should not be served in prison. Hospitals caring for such persons would, after all, be prisons in which the addicts, not already cured when they arrive, would be in bed six days more or less and in a ward or dormitory resting for two or three weeks. The Federal prisons are already equipped for this sort of treatment.[57]

In 1928, Kolb ended his assignment with the Hygienic Laboratory and was sent to Europe for several years to evaluate methods of screening for intelligence and mental fitness in prospective immigrants. Upon his return to the United States in 1932, he was put in charge of planning and managing a Public Health Service hospital for defective delinquents in Springfield, Missouri. Meanwhile, his view that prison hospitals for addicts were unnecessary failed to carry the day.

6

Healing Vision and Bureaucratic Reality

To criminalize opiate addiction with the stroke of a pen was one thing; to process and manage addicts as they were swept up in arrest and moved through the court system and into jail or prison was quite another. Addicted prisoners presented a problem to the law enforcement system at every jurisdictional level. If actively addicted at the time of arrest, they went into withdrawal in jail cells. If imprisoned for any length of time, they were likely to try, often with success, to smuggle drugs into their jails or prisons. Frequent recidivism made nuisances of addicts brought repeatedly to courts for arraignment or sentencing. Courts sometimes attempted to offload the problem by sending addicts to state institutions, such as the California State Narcotic Hospital at Spadra, but managers of these institutions complained that these seasoned addicts undermined the therapeutic mission of hospitals designed to treat curable (that is, earlier-stage) addicts.[1]

At the federal level, enforcement of the Harrison Narcotic Act had similarly affected the inmate population. In April 1928, of a total of 7,598 federal prisoners, 1,600 were addicted to opiates.[2] The wardens of the three federal prisons (including Leavenworth, where Clifton Himmelsbach had developed his addictiveness assay) did not share Lawrence Kolb's view that addicted prisoners could be adequately detoxified and housed in prisons. In 1928, when Congressman Stephen Porter (R-Pa.) introduced legislation to authorize the creation of two narcotic hospitals to house addicted federal prisoners and probationers (as well as voluntary patients), federal prison wardens were among its supporters.

The Public Health Service Narcotic Hospital at Lexington, Kentucky,[3] and its companion institution at Fort Worth, Texas, were intended from the outset to function as both prisons and hospitals. An institution that doubled as a hospital and a prison was no anomaly, given the long shared history in Europe and America of these two kinds of institution of confinement. The hospital and the prison show contrasting faces of societies' sometimes blurred distinctions between dependency that merits pity and conduct that deserves incarceration. The *hôpital général* of early modern France, for example, was an institution of last resort where the state immersed idlers, vagrants, and the unemployed poor in work and religion to make them more productive subjects in a mercantilist monarchy.[4]

More immediate antecedents for the PHS narcotic hospitals included the nineteenth-century asylum for the mentally ill, discussed in Chapter 5, and the inebriate sanitarium. Inebriate sanitaria dating from the late nineteenth century provided treatment for those suffering from inebriety, that is, dependence on alcohol or opiates. By the mid 1880s, prolonged sequestration in a rural setting was the standard treatment for inebriety.[5] Meeting a broad market demand, sanitaria for inebriates varied widely in quality and method in their treatment of opiate addiction. Some, like the Montefiore Sanatorium in Westchester County, New York, included farmwork in their therapy.[6] The worst simply provided opiates under the guise of treatment. Some offered little more than a retreat for managing detoxification and an indefinite stay in a low-stress setting with rest and nutritious meals. Many deployed specific treatments for opiate addiction, ranging from quack remedies to management based on the latest medical modes. However variable their quality, they met a real demand for relief from opiate dependence. In the 1920s, as prohibition of both alcohol and opiates undermined the mission of these institutions, their numbers decreased substantially.[7]

Both insane asylums and inebriety sanitaria provided precedents for institutionalizing addicts (defined as suffering from mental illness) at a time when addicts were being imprisoned for possession (and sometimes sale) of drugs. Criminal justice management of addicts created the context for defining addiction not just as a diffuse disorder (as reflected in Chapters 4 and 5) but as a serious form of mental illness, comparable to schizophrenia or incapacitating depression, in that it warranted lengthy, forced institutionalization.

When Stephen Porter brought a bill before Congress in 1928 to create a new kind of institution, to be called "narcotic farms," its supporters included both law enforcement officials and scientific experts on addiction. Federal penitentiary wardens saw addicts as troublesome. The Justice Department wanted a better sentencing alternative for those convicted under the Harrison Act. Regular prison sentences seemed unduly harsh in some cases. By the late 1920s, those experienced in addiction treatment recognized the frequency of relapse following withdrawal, and the drastic solution of confining addicts was consistent with the prevailing therapeutic pessimism. Scientists interested in addiction approved the inclusion of research facilities.

In envisaging new prisons whose functions would be rehabilitative, not simply punitive, American prison reformers of the 1920s such as Thomas Mott Osborne and Frank Tannenbaum looked to the hospital approach of diagnosis, triage, and individualized treatment as a model.[8] The prison should be structured as a community within which the prisoners would learn to be good citizens. Public health analogies suggested that the prison was like a hospital that offered a form of quarantine against the diseases of society. A medical model suggested that prisoners, like patients, should be diagnosed, classified, and triaged into the best rehabilitative environment.[9]

Porter's bill became law in 1929. It authorized the establishment of two narcotic farms, one in Lexington, Kentucky, to house addicts from east of the Mississippi and one in Fort Worth, Texas, to receive addicts from west of the Mississippi. Both institutions were intended from the outset to function as federal prisons as well as hospitals for the treatment of addiction. Lexington's planners, like the legislators who passed the Porter Act, saw the institution as part of a wave of prison reform that would rehabilitate prisoners rather than simply locking them up. They expected to minimize recidivism through evaluation of incoming inmates, triage to appropriate wards and behavioral regimens, and social work to guide inmates back into the outside world.[10]

Lexington was also a place where psychiatry would be showcased. The first medical director was the PHS psychiatrist Lawrence Kolb, and the institution, as a Public Health Service hospital, trained psychiatric residents. The Public Health Service Narcotic Hospitals were planned and constructed at a time when American psychiatrists (still largely an asylum-based specialty in spite of the growth of urban psychopathic hospitals) were rediscovering the ideas of early nineteenth-century moral therapy and reemphasizing the importance of molding hospital routine as a treatment method. Energized by the new psychiatric ideology of adjustment, asylum superintendents and staff psychiatrists saw their institutions as sites where treatment would consist of changing patient behavior so as to promote readjustment to the social demands of the outside world. As William Russell, the president of the American Psychiatry Association, said in 1932, the psychiatrist's most important therapeutic resource was the social world of the hospital itself.[11]

The two planned Public Health Service narcotics hospitals were to be managed by the Division of Mental Hygiene, created in 1930 as authorized by the 1929 Porter Act, for this purpose. (The act called for a Narcotics Division in the PHS, and an act of June 14, 1930, changed the name to the Division of Mental Hygiene.)[12] Walter L. Treadway, another Public Health Service psychiatrist, was named to head the division (in this role, he oversaw clinical testing of compounds produced by Lyndon Small's group at the University of Virginia, as described in Chapter 3). Treadway, like Kolb, had a career in public health psychiatry. He had worked under Thomas Salmon at Ellis Island screening prospective immigrants for mental fitness and had conducted field studies on the efficiency of local social services and on the comparative health and mental status of native-born versus immigrant groups. In 1922, he had been named medical officer in charge of the newly created PHS Office for Field Studies of Mental Hygiene, based in the Department of Prevention Medicine at Harvard Medical School.[13]

Treadway envisaged the institution as a hospital rather than a prison. His correspondence with John D. Farnham, an attorney engaged in prison reform work for the Bureau of Social Hygiene in the early 1930s, reveals his plans for Lexington. He believed that the length of time that prisoners and probationers would spend at the hospitals would make rehabilitation possible. There would be no surrounding walls. Amenities would include a gymnasium, an auditorium, and facilities for tennis, baseball, and other outdoor sports. To accomplish the rehabilitation, Treadway had in mind at least five regimens, each with its own combination of housing, diet, and treatment components. Inmates should arrive with complete social histories, which would inform treatment planning. Following an inmate's release, a social service staff would plan reintegration into the community—not necessarily in the place where the addict had acquired his or her bad habits, but perhaps somewhere new, where "Rotary Clubs or other organizations of similar nature" might assist in finding appropriate jobs.[14] Treadway's selection of the middle-class and entrepreneurial Rotarians as appropriate mentors for newly released, mostly working-class ex-addicts echoes Roy B. Richardson's exhortations that his patient cultivate contacts with "decent people." Treadway's expectation that Rotarians would adopt newly released addict-prisoners was undoubtedly overly sanguine.

Just as he had seen the project to develop a nonaddicting analgesic as connected to the federal policy of drug criminalization (because the improved medication would eliminate the possibility of iatrogenic addiction), so he saw Lexington's rehabilitative mission as incorporating the law enforcement function into the new vision of the prison as a site of rehabilitation. He and Farnham presented Commissioner of Narcotics Harry Anslinger with a plan to have Anslinger's field agents collect histories from their arrestees before starting the court procedures that would presumably bring them to Lexington. Such histories, in Treadway's view, would provide important input for treatment planning.

For Treadway, the institution's objectives were to restore its inmates to health, to rehabilitate them, and to train them to be self-supporting. Vocational training and education would accomplish the latter result. At the same time, custodial features would be emphasized, because the institution was intended specifically for "the care of the more intractable type of persons."[15]

The newly built Public Health Service Narcotic Farm (as the institution was named in the legislation) was dedicated in a ceremony on May 25, 1935, just outside of Lexington, Kentucky. Surgeon General Hugh S. Cumming was the main speaker. The opening of the hospital, Cumming noted, exemplified the enormous expansion in the scope of Public Health Service activities since he had been commissioned forty-one years earlier. Then, the Public Health Service had only cared for the sick and wounded in the military and the merchant marine. (In fact, Cumming's historical memory was faulty; by 1894, the Marine Hospital Service was already engaged in immigrant screening and quarantine work, and its Hygienic Laboratory had been functioning for seven years.) Now, Cumming said, PHS responsibilities included control of food and drugs, assistance to state public health programs, and, increasingly since World War I, the medical treatment of individuals. The inauguration of a new institutional approach to the care of narcotics addicts marked a recognition that "restrictive laws governing commerce in narcotics" were not the only necessary policy thrust in dealing with addiction.

Cumming described addiction in public health terms, as if it were a contagious disease. As the target of the illicit market in opiates, addicts were in essence a source of contagion, endangering their fellow citizens. Addiction spread through contact between addicts and other people. Therefore, segregating addicts from society "with the object of medical treatment" would protect the public as well as helping the addict. Segregation alone was not an adequate solution, however, since addicts had become the most common form of repeat prisoner among federal offenders. Moreover, he noted, addiction to habit-forming drugs occurred in every segment of society. No factor such as age, sex, occupation, or nativity exempted one from liability to addiction; thus, addiction resembled "an endemic disease" warranting a "medico-social" response. Cumming compared the treatment of addicts to the treatment of the insane: in past, less enlightened periods, simple confinement of the insane had prevailed, but medical progress had now provided humane, therapeutic regimens for those suffering from mental illness. Similarly, the Lexington Narcotic Farm formed a part of society's progressive approach to the problem of aiding those who became public charges. Addicts had failed to meet the challenges of an increasingly complex civilization, and the mission of Lexington would be to bring important medical advances to bear on this problem, which enforcement alone had not ameliorated.[16]

These early visions of Lexington reflect the initial hopes for an institution that later came to symbolize the intractability of addiction linked to the urban underworld. They also represent a melange of concerns underlying an attempt to make society's response to addiction more modern, more humane, and more effective. But the policy structures and reform currents framing both public and philanthropic responses to addiction in the 1920s constrained scientific and medical research on addiction in this period to approaches consistent with social control of addicts.[17] The nation's commitment to control of worldwide drug supplies and enforcement of prohibitionistic drug legislation, as well as a general hardening of Progressive reform impulses into systems of classification and segregation, limited the range of acceptable solutions to the addiction problem. The networks of personal connections among leading physicians and scientists, officers of philanthropic foundations, and government officials further encouraged the development of research and policy initiatives, which, if not always mutually reinforcing, were at least compatible or not mutually contradictory.

As a "narcotic farm" with up-to-date research facilities for clinical drug testing, Lexington blended aspects of the past and the present. Cumming's dedicatory remarks similarly reflect traditional and new thoughts about larger issues of caring for society's dependents and more specific problems associated with drug use. In citing the history of the mental asylum, Cumming placed Lexington in the prevailing legend of psychiatry's own history: a bold story of scientific advances vanquishing old superstitions, of humane medical treatment of the mentally ill replacing stark and cruel confinement. Comparing the Narcotic Hospital to an asylum for the dependent insane also placed the newer institution in the centuries-old tradition of charitable care for the needy. Implicitly, Cumming portrayed Lexington as a modern, scientific way for a community to care for troubled individuals.

Cumming's description of addicts as both market consumers of drugs and contagious agents of addiction touched on the capacity of public health policy to embrace both medical and enforcement concerns. Protection of the public underlay the coercive power of health authorities to intervene in the private lives of individuals; if addicts spread their disease to others, then separating them from society was implicitly justified. This link also reflects the consistency between the supply-side enforcement policies pursued by criminologists like Lawrence Dunham of the Bureau of Social Hygiene and by bodies like the Federal Bureau of Narcotics, on the one hand, and medical treatment of addicts, on the other. Cumming also expressed the medical view of social deviance (embodied in psychiatric theories about adjustment) when he described addicts as people who had failed to meet the challenges of an increasingly complex (that is, industrial and urban) American society.

Walter Treadway's statement of Lexington's objectives focused more directly on the hope that a variant of the prison could provide a form of treatment that helped failed individuals take up productive social roles upon release. That an institution designed to prepare rehabilitated addicts for productive life in an increasingly industrial and urban society was initially envisaged as a farm suggests the juxtaposition of a perception of modern social problems with an axiomatic sense that the traditional values of America's agricultural past were a model of personal and civic health. It also reflects psychiatrists' renewed interest in moral therapy and the Kirkbridian asylum in the early 1930s. The vocational activities developed for the inmates actually included woodworking, furniture making, and garment manufacture and repair, in addition to an array of food-production activities: raising and slaughtering pigs and processing the meat; managing a dairy herd and hennery; raising and processing a variety of crops.[18]

These plans reveal an institution modeled on prisons that included farmwork as rehabilitation and on mental asylums. Several factors made asylums function much like prisons. In the 1930s, two trends converged to intensify the social control aspects of asylum life for inmates. First, in the late 1920s and early 1930s, "moral therapy," a treatment philosophy and regimen developed in the early nineteenth century, enjoyed a renewed vogue.[19] Moral therapy had provided some of the inspiration for Thomas Kirkbride's model of the rural asylum. The central idea of moral therapy was that a well-ordered and highly structured life in the institution constituted its main therapeutic impact on the patient. As discussed in Chapter 5, this idea meant that institutional administration and medical functions were combined in a single enterprise.

Second, asylum superintendents reembraced moral therapy just as a new set of shock therapies were being developed. As psychiatrists became disillusioned with the ability of drugs like hyoscine to calm patients, they turned to increasingly invasive somatic therapies to control patients' disturbed behavior. These somatic therapies, introduced in the 1930s, included shock therapies based on inducing high fevers or insulin shock and culminated in lobotomy.[20] As Jack Pressman has shown, asylum superintendents deployed invasive treatments like psychosurgery as tools to facilitate managing patient behavior in crowded asylums.[21]

Similarly, Joel Braslow has examined patterns of therapeutic intervention in California asylums and argued that in these institutions, behavioral control and medical management were seamlessly intertwined. Braslow argues that hydrotherapy, the practice of wrapping patients tightly in sheets and soaking them for long periods in warm baths, marked a transition toward the deployment of increasingly severe therapies as methods of behavioral control. The main impact of these combined trends for inmates was that every aspect of life in the asylum, from the blandest daily routines to the most invasive therapies, was organized to control inmate behavior and keep it orderly. From the perspective of the psychiatrists running the asylum, every aspect of life in the asylum represented a component of therapy that they believed would restore patients to the maximum level of social functioning they were capable of. Treadway's recommendation that addicts would benefit from hydrotherapy further indicates that he framed chronic opiate addiction as a form of mental illness comparable to the illnesses suffered by asylum inmates.

Lawrence Kolb's appointment as the hospital's first medical director was ironic, given his argument against the need for such institutions. In the months preceding the opening of the institution, Kolb both planned the treatment regimen and directly oversaw every phase of the hospital's construction and furnishing. Requisition forms for everything from beds to stationery to seed stock for the farm bear his signature.[22]

Kolb saw the hospital as an alternative to prison for addicts. Seven months after the hospital opening, he wrote: "The public is served through this agency by a more human and understanding treatment of convicted drug addicts than was possible when such addicts were sent to penitentiaries and treated as criminals. Stigmatizing these people as criminals for what in a very large proportion of cases was merely a weakness had the contrary effect to treatment."[23] He went on to say that the hospital would provide a needed treatment resource for individuals voluntarily seeking treatment, "as it is a fact that throughout the country hospitals as a rule will not receive drug addicts, or will not keep them long enough to effect a permanent cure and the sanitaria that do receive such patients necessarily charge sums beyond which the average addicts is able to pay." Finally, since addicts "act as a source for the spread of addiction to others," it was important to hospitalize them for long enough to effect a meaningful change in their behavior. He conceded that, although the "institution is operated as a hospital . . . it necessarily has prison features."

From the outset, Lexington was also envisaged as a research site. All the inmates were addicted to opiates (and, until a women's section was opened in the 1940s, all were men). They included prisoners with terms to serve, probationers whose condition of release was to be pronounced cured by the hospital staff, and voluntary admissions, who were free to leave whenever they wanted, even if against medical advice.

Kolb planned a regimen suitable for the psychopathic addicts he had described in 1925. The treatment regimen was divided into three phases.[24] First came the withdrawal phase. This began with stabilization of the addicted inmate on the lowest dose of morphine that would prevent withdrawal symptoms—in other words, an amount approximately equal to the amount of morphine or heroin the addict had been using before entering the hospital. This step enabled the staff to determine the severity of the inmate's addiction, which helped chart the course of treatment and was also important for research into the physiological nature of addiction. Following stabilization, the morphine was abruptly withdrawn and the inmate suffered through the abstinence syndrome. This phase was typically over within ten days.

The initial treatment phase also included information gathering and evaluation. The patient was interviewed, and the social service staff contacted agencies, relatives, and others who knew the patient to assemble a profile. Sources besides the patient were necessary to corroborate the patient's story, which was not deemed reliable by itself. At the end of thirty days, the hospital's classification board considered the patient's case and determined his classification according to Kolb's six-part typology.

Phase two of treatment took up most of the remainder of the patient's stay at Lexington. Kolb characterized it as "building up the patient physically and mentally"[25] and recommended a minimum stay of six months for voluntary patients. The entire routine of life in the hospital was designed to provide a healing environment. Each patient had a work assignment, which was considered "vocational training." In actuality, these were not training programs designed to equip the inmates to take up specific kinds of jobs upon leaving the institution, but were kinds of work that reflected the melding of therapeutic and administrative aspects of institutional life. Some jobs involved agriculture and animal husbandry on the institution's farm; these were specifically intended to have a therapeutic effect. Others involved making furniture or clothing; the products of these efforts would be distributed to other prisons in the growing federal prison system. Finally, some jobs, like working in the laundry, contributed directly to the functioning of Lexington itself. Other aspects of life such as bedtimes, meals, and recreation were also routinized.

In sum, life at Lexington would resemble life at asylums for the mentally ill and life at other American prisons. Work was understood to be therapeutic even as it contributed to the running of the institution. Belief in the rehabilitative value of work, including farmwork, was also evidenced in some of the new prisons being built around the same time. For example, in 1919, the state of California began sending women convicted of prostitution to the California Industrial Farm for Women. In 1927, the first federal prison for women, the Federal Industrial Institute for Women, was opened at Alderson, West Virginia. Here, women engaged in farmwork and in vocational training, including typing and filing, which reflected gendered expectations of upward mobility into respectability for women. By 1935, the year Lexington opened, over half of Alderson's inmates had been sent there following arrest on narcotics charges.[26] (The salience of prostitution and narcotics violations in driving up the number of female prisoners points up the impact of Progressive Era vice reforms in creating new classes of criminals.)[27]

During their stay at Lexington, inmates benefited from medical and dental services, which addressed complications of addiction as well as other health concerns. During September 1935, when there were almost 600 inmates, the most frequent surgical procedures were drainage of skin infections, common results of faulty injection technique; but almost as many procedures were for excision of ingrown toenails. At least a sixth of the inmates received vaccinations for smallpox, typhoid, or other diseases. Venereal disease was treated in 723 cases, but it is impossible to know how many inmates were involved (treating syphilis then typically involved multiple doses of arsenicals and bismuth). The records also note 382 psychiatric treatments, but without further details, and, again, it is impossible to know how many patients were involved. Psychological tests such as the Army Alpha test (precursor to the IQ tests that became standard in the schools during the 1920s), or tests like those in which inmates were called upon to match shapes, were given to 48 patients.

By far the most common form of treatment was dental: 1,295 treatments for about 600 patients were recorded in the single month. Addicts were well known to have serious tooth decay problems; reports on patients in the narcotics ward at Philadelphia General Hospital frequently refer to mouths filled with little more than blackened stumps. The causes were poor nutrition and prevalence of refined sugar in the diet, as well as general neglect of health, all patterns associated with heavy addiction. However, it is likely that dental caries were common among working-class adults in the 1930s, and that these patients had had little dental care in their lives. Access to a dentist and to the other medical services offered at Lexington were thus significant benefits for the inmates.[28]

The third phase of treatment involved preparing the patient to return to the world outside. This phase was crucial, because relapse rates were known to be high, and the role of the environment in triggering relapse was considered a point of particular jeopardy. The objective of this third phase was to attempt, through the social service staff at the hospital, to facilitate the patient's transition into an appropriate employment and residential setting. Ideally, there would be a good job and welcoming family waiting for the patient. The importance of vocational training, such as it was, lay in its preparation of the patient to secure gainful, legal employment following release.

The unifying theme in the second and third phases of treatment was adjustment. Patients had become addicted because of social maladjustment, and the goal of treatment was to prepare them to make a better social fit after release. They learned to do this by adjusting properly to the simulated community in the hospital. Thus, treatment would include "adjustment of outside conditions [e.g., family reception or employment placement] in preparation for his eventual return to society"; withdrawal was discussed in terms of "physical readjustment"; and treatment's "ultimate aim . . . [was] to bring about a health adjustment to life without drugs." Similarly, Kolb's description of the treatment regimen continued, "more helpful habits of adjustment are being formed through work and controlled recreation"; "requiring certain lines of conduct and behavior that are necessary for the orderly adjustment of the patients to one another and to the hospital staff is not regarded by the patients as an unreasonable coercion"; "general policy is to regard all infractions of discipline as mental maladjustment and to find a solution that will bring about a satisfactory readjustment."[29]

Administrative records for the hospital's operation from its opening in 1935 to 1940 reveal tensions that characterized the early phase of attempting to combine prison and hospital facilities for the amelioration of addiction. Two areas of concern reveal inherent contradictions involved in viewing addicts simultaneously as prisoners, patients, and future effective citizens. These consist of the role of occupational therapy and education in a highly bureaucratized and authoritarian setting, and the problems involved in adapting the institution to the needs of prisoners with fixed sentences, probationers whose length of stay was at the discretion of the medical personnel, and voluntary admissions. A 1939 analysis of the inmate population by the Public Health Service psychiatrist Michael Pescor offered further support for hardening professional perceptions of the intractability of addiction. The coercive qualities inherent in Lexington's institutional structure—resulting from its essential character as a prison, from its role in a massive federal bureaucracy, and from its therapeutic assumptions about the social defectiveness of its inmates—soon hardened into an expression of profound pessimism about the possibilities of reclaiming addicts as productive and acceptable members of society.

A Healing Farm in a Bureaucratic Matrix

Lawrence Kolb's typology of opiate addicts influenced how inmates at Lexington were sorted into treatment regimens, and Lexington replaced the Prison Annex at Fort Leavenworth, Kansas, as the site for the clinical studies of test drugs developed under the auspices of the National Research Council's Committee on Drug Addiction. Lexington was thus the institutional site where various strands of addiction research and policy came together, and where it was expected that important solutions to the addiction problem would emerge.

The anticipated date for first receipt of inmates was pushed back several times from February to July 1935; in the meantime, institutional staff planted that year's crops and established the husbandry activities. As medical director of an institution in which these activities constituted a component of therapy, Lawrence Kolb was involved in setting up procedures for overseeing the various productive activities. This involved weaving the farm and shop work into a bureaucratic mesh that connected details of daily life at Lexington to the highest levels of oversight in Washington, D.C.

Multiple layers of authority overlaid the social space to be inhabited by Lexington's inmates. The most obvious level of coercion was embodied in the custodial features mentioned by Treadway: the institution's character as a prison. Less overtly authoritarian, but equally pervasive in the routinized daily life at Lexington, was the medical authority that triaged inmates into diagnostic categories and involved them in activities believed to be curative of their fundamental personality disorders. As in mental asylums, Lexington's patients were treated much like prisoners. For example, mental patients were transferred between institutions in handcuffs; within the asylum, their mail was censored, they had no privacy, and their personal belongings were subject to search at any time.[30]

The Public Health Service, as Cumming noted, had grown into a bureaucracy of massive proportions by the mid 1930s. It was managed in a top-down authoritarian style, fashioned like a military chain of command, with commissioned officers working in a hierarchical structure. The penetration of bureaucratic procedures into the therapeutic realm of the farm and other vocational activities is revealed in correspondence between Kolb at Lexington and Walter Treadway at the Division of Mental Hygiene in Washington over such matters as establishing record-keeping and accounting procedures for Lexington.

A request to the Department of Agriculture for copies of forms useful in maintaining a dairy herd prompted a reply recommending several from among the large number of forms used in connection with the Bureau of Dairy Industry's own herd. These included one form per cow for tracking daily milk production and two forms comprising a permanent record for each cow.[31] Kolb planned a series of accounting records to track pigs from the piggery where they were raised to the slaughterhouse to the cannery. First the animals and then their meat became assets charged against or credited to each of these units as they moved through each step of raising and processing. Since the resulting food and other products (e.g., hides) could either be used at the prison or sold, additional accounting procedures reflected the gain to the institution from its various productive activities. Other accounting data included daily reports of fuel use by tractors and other farm vehicles. Even the manure produced by the dairy cows and hens was treated as an asset, credited to these operations, and charged against the grain farm, truck farm, or greenhouse operations where it was used as fertilizer.[32]

The productive activities were not treated as freestanding businesses; any profit or loss in the various activity sectors was incorporated into the overall budget of the institution. Moreover, products from the manufacturing shops (such as the garment shop) were intended for use at other institutions in the prison system. Thus, the economic activities of the inmates, although rhetorically and therapeutically intended to promote a capacity for well-adjusted self-sufficiency, were, in fact, blended seamlessly into the workings of a large bureaucracy, whose management functions aimed at fiscal efficiency in a vast institutional network. As Surgeon General Cumming commented in a communication to the secretary of the Treasury in seeking approval for manufacturing goods at Lexington: "The establishment of these proposed activities bears a direct relationship to therapeutic occupation of inmates and the partial liquidation of costs involved in their maintenance and care."[33]

Lawrence Kolb was minutely involved in developing these procedures. For example, in correspondence with Treadway (who from his office in Washington was able to veto any proposal if it violated some aspect of PHS or other federal regulations), Kolb made specific recommendations about the accounting methods for the various areas of operation, including what prices to charge for the various commodities produced.[34] Kolb saw no inconsistency between the therapeutic aims of the institution and its functioning as a component in an extensive bureaucracy managed from the top down. Although the therapeutic regimen was described in the terms of the new psychiatry of adjustment, many of these institutional practices were holdovers from the nineteenth century. As noted in Chapter 5, a conflation of therapeutic and administrative concerns, both in patient activities and in the superintendent's responsibilities, characterized the nineteenth-century insane asylum. Similarly, putting inmates to work as a means of recovering some institutional operating costs of prisons was common in the nineteenth century. Such practices were discontinued when prisons were recast as rehabilitative institutions, but at Lexington, the expected therapeutic value of work justified activities that would be remunerative for the institution.

The range of job functions available for "occupational therapy" did not extend beyond the functions involved in running the hospital/prison and the farm and producing garments and furniture for sale to other parts of the prison system. Occupational therapy was divided into maintenance (including cleaning and food service), trades (including laundry, woodworking and machine shop) and farmwork (including field work and animal husbandry).[35] Undoubtedly, some inmates learned useful skills in some of these positions, especially those in furniture-making and machine shops.

However, the main purpose of the occupational therapy was to adjust attitudes, not to impart skills. Lexington's managers wanted to improve inmates' capacity to "adjust," to fit into their appropriate social roles. A number of factors, not least the preponderance of working-class men in the inmate group, supported casting most addicts' appropriate place in the social structure as a modest one. Kolb had characterized addicts as men whose social and hedonistic ambitions outstripped their narrow abilities and humdrum prospects. Roy B. Richardson's routine labeling of addicts as "constitutional inferiors" paralleled this view. Mental tests provided a seemingly precise means of determining individuals' capacities; on the basis of intelligence tests, the first 1,036 admissions to Lexington were estimated to have an average mental age of 13 years and 8 months (with 15 years considered average).[36]

To a degree, Kolb and the physicians and researchers at Lexington were correct: improved skills alone would not have resolved the problem of addiction; moreover, no amount of job training could compensate for the unemployment effects of the Depression. Virtually all the inmates at Lexington in September 1935 reported occupations that involved some degree of skill.[37] Lexington's managers rightly focused on the addict's problems in sticking to a job. Their proposed solution, however, was to have the inmates live in a simulated community where they would develop positive habits of routine and work, with the expectation that such habits would improve clean living and regular work habits back home. Implicit in this system was the view that addicts needed to reconcile themselves to lesser social roles than they had envisaged.

Early surveys of the Lexington population make it clear that most inmates lived in deteriorated urban centers.[38] Farmwork would hardly prepare them to resume city life in any direct way; the impact of institutional routine on character was the intended benefit. (The greatest actual benefit for malnourished, underweight, post-withdrawal addicts may have been outdoor exercise, along with regular meals eaten with a farmworker's appetite.) The various forms of indoor work, ranging from pushing brooms and folding laundry to making furniture along prescribed institutional lines, resembled more closely the kinds of work inmates had come from and would likely return to.

The relationship of addiction to occupation can also be considered in terms of the interaction between work and addiction management. First, managing an addiction is a burden that one's livelihood must somehow support. One advantage of viewing addiction as a chronic illness is that this perspective highlights the kinds of work that go into managing the addiction, including acquiring the resources to buy the drug and controlling unwanted side effects like constipation. The work involved in hiding one's addiction and passing as a nonaddict also has parallels to highly stigmatized chronic diseases. Like chronic illness, addiction could be the crucial factor that overwhelmed an individual's capacity for self-sustenance. As some of the cases described in Chapters 1 and 4 illustrate, working-class wages were often insufficient to meet the cost of drugs on top of such requirements as food and shelter. At all class levels, an addiction to morphine or heroin could get out of control and take over a life, but for those living near the margin of subsistence, the cost of drugs, determined by the illicit market, could exceed earning capacity even when the dose was kept fairly steady.

Second, some occupations pose particular risk for addiction to opiates and some occupations lend themselves better than others to addiction management. Examined in this light, the range of occupations reported by the 563 inmates at Lexington at the beginning of September 1935 is illuminating.[39] Two work sectors, entertainment and health care, could involve occupational exposure to opiates. This risk has long been recognized for health care workers who handle medicines. The thirty-two health care workers among this group included twenty physicians; by contrast, only three other individuals of comparable occupational status, one attorney and two engineers, were Lexington inmates during that month, along with four dentists, two nurses, one osteopath, and one drug clerk. Ninety-four inmates (approximately 17 percent) had worked in occupations connected in some way to the nexus of entertainment, amusement, and vice activities that flourished in urban centers. These included seventeen chauffeurs. ("Chauffeur" or "driver" was a common euphemism for a gang henchman, whose duties were not restricted to driving; but even private chauffeurs for legitimate employers would have to know their way around a city and would have idle time to spend in street venues while waiting for their passengers.) In addition, there were fifteen cooks, twelve waiters, nine porters, five hotel employees, and five bartenders. Four gave their occupation as gambler; three as actor, musician, or racetrack worker, and two as bellhop, poolroom operator, or horse trainer. The remainder represented such occupations as fight manager, taxi driver, and carnival concessionaire. Together, occupations linked to health care or amusement accounted for just over 22 percent of the inmates.

Having an opiate addiction militated against long-term employment in various ways. Arrests, jail time, lengthy cures, or exposure as an addict could force one out of a job. Drug effects such as nodding off could undermine job performance. Crises of drug availability could send one frantically in search of a supplier while withdrawal symptoms were coming on; a succession of these might create a pattern of tardiness or absence. The common sequence of addiction, cure, abstinence, and readdiction could be paralleled with cycles of employment, loss of job, restabilization, and renewed employment.

These factors would all favor kinds of work that were episodic in nature, or that one could move into and out of with relative ease. For example, day laborers could move in and out of the job market on a short-term basis. Various kinds of construction and contract work involved a succession of short-term job commitments. Work in which one controlled one's movements could afford chances during the day to administer drugs in private; salesmen who called on accounts or sold goods from door to door or on the street had this advantage. Forty-six of the inmates gave salesman as their occupation, and the two demonstrators, three hucksters, two newsboys, and a single junk dealer in the group would have shared the same advantage. Numbering 57, those reporting some sort of sales occupation represented just about 10 percent of the inmates. The Lexington group included an embalmer, two crane operators, and a piano maker; workers with special skills like these might be able to count on employment, perhaps with a new employer, following bouts of prison or cure.

At the same time, many of the occupations suggest no obvious link to opiate addiction; rather, their variety indicates that opiate addiction could be, and was, woven into many ways of life. The illicit market could only survive with a customer base that extended well beyond the immediate nexus of vice and entertainment, and the range of kinds of work the inmates reported supports that view.

It is impossible to know how central the reported occupation was for any given individual. One might report the legitimate side of a life that also involved illicit means of earning money, or one might report one's last legitimate employment before relying completely on illicit earnings. Comparing the reported occupation to the life narrative for cases at Philadelphia General Hospital revealed many such instances. Nevertheless, the overall picture suggested by the list of occupations includes several salient features. There are definite clusters in areas with clear likelihood of exposure to opiates or with some advantage regarding habit management. At the same time, most reported occupations had no such link. As might be expected, the great preponderance were working-class, broadly defined. They included unskilled laborers and factory workers; semi-skilled workers such as housepainters (16), railroad workers (9), laundry workers and mechanic's helpers (one each); and skilled workers, or tradesmen, such as machinists (10), tailors (2) or welders (1), electricians (8), barbers (18), and plumbers (3). The thirty-nine clerks are not easy to classify; they may have worked in retail establishments or in offices. Twenty-three indicated they were in business for themselves; seven white-collar workers included four bookkeepers and an accountant, a reporter, and an advertising man. Nineteen identified themselves as farmers. It is likely that middle-class and professional groups were slightly underrepresented compared to their actual prevalence among opiate addicts, because they typically had resources that sheltered them from some of the risks of both addiction and the illicit market. The case of the accountant Everett Lewis (discussed in Chapter 4) illustrates such factors.

Prisoners, Probationers, and Voluntary Patients

Given Lexington's status as a prison, therapeutic considerations frequently gave way to other imperatives. Ideally, from Kolb's point of view, medical considerations would dictate the length of inmate stays at Lexington. He believed six months to be the necessary minimum. In fact, various types of inmates were kept under differing legal conditions, which interfered with purely medical judgment about length of stay from both directions: prisoners might have sentences requiring stays longer than the medical personnel might deem necessary, while voluntary patients could leave at any time.

One visible tension was that between Kolb's desire to optimize the prospects of success for incoming patients and the Bureau of Prisons' intention to offload as many already incarcerated addicts as possible to the new institution. Lexington's planned capacity was 1,200 beds, some of which would be in specialized wards such as a tuberculosis ward and thus not open to healthy arriving inmates. In 1932, while Lexington was still under construction, Walter Treadway noted that there were currently at least 1,500 drug addicts in federal prisons.[40] Most of these were being housed in the Army Disciplinary Barracks at Fort Leavenworth, Kansas, which the Army had made available to the Bureau of Prisons for this purpose pending the opening of Lexington. Soon after Lexington began accepting prisoners, Sanford Bates, director of the Bureau of Prisons, communicated a desire to transfer prisoners from Leavenworth to Lexington in batches of fifty until the latter facility was filled.[41] To fill Lexington with prisoners already serving sentences, and already withdrawn from addicting drugs, would frustrate several institutional aims, however. Actively addicted individuals were necessary for Clifton Himmelsbach to continue his clinical studies with the test drugs developed under the auspices of the NRC's Committee on Drug Addiction.

At the Bureau of Narcotics, Harry Anslinger was polling judges on their views regarding sentencing addicts to Lexington and instructing his field officers to direct addicted individuals to Lexington for voluntary admission. He also asked Treadway for a list of all known addicts so that he could mail them a circular urging them to enter Lexington of their own accord as patients. Finally, he anticipated an increase in addicted prisoners coming from the court system. He believed that his efforts against large-scale drug traffickers had been so successful that he would soon be able to direct his bureau's enforcement efforts against retail-level peddlers, who, unlike the wholesalers, were frequently addicted themselves. Sentencing such individuals to Lexington, he noted, would be doubly effective, simultaneously reducing the number of both consumers and sellers in the illicit drug market.[42]

Kolb, meanwhile, expressed the hope that all inmates sent by the courts would be probationers, as this status would allow the medical staff to determine when they were sufficiently improved to warrant release. Similarly, Kolb felt that voluntary admissions should be restrained from leaving against medical advice if at all possible. Lexington was a prison, but early discussions quickly established that firearms or potentially injurious physical force would never be used to prevent either voluntary patients or probationers from escaping. Finally, as busloads of transfer prisoners from Leavenworth threatened to fill Lexington with already convicted, already withdrawn addicts, Kolb expressed concern that Himmelsbach's clinical studies could not proceed without a sufficient supply of inmates (whether from the courts or as voluntary admissions) not yet withdrawn from opiates.

Lexington's classification board, as noted above, made treatment recommendations for each inmate, following an initial observation period, according to Kolb's psychiatric categories of addiction etiology. The variety of legal statuses of different groups of inmates, however, added another dimension to the sorting of inmates, and the classification board became involved in these issues as well. It decided that voluntary patients should be housed separately from prisoners or probationers, and the level of surveillance applied to these patients was substantially less rigorous than that deployed in other wards. The number of voluntary patients in the initial months was very small. Any illusion that the treatment regimen was pleasant or compelling enough to reliably retain voluntary patients was shattered when four of them went over the wall one night in August, the third month of Lexington's operation.

Attendants learned of the escape from a physician inmate. Shortly afterward, one of the escapees, who had sprained an ankle during the escape and sought help, was found and quickly returned to the institution. The other three were not found, and, after an initial search of the grounds and the immediate vicinity, no attempt was made to recapture them. Had they apprehended the escapees, staff had no legal grounds for compelling their return. Attendant-guards recounted their instructions regarding surveillance of voluntary patients and described their routine in this portion of the institution. Guard staff were still somewhat shorthanded, and only one day attendant kept watch on the voluntary group. During the night, periodic checks replaced constant attendance. The wall surrounding the recreational courtyard made available to the voluntary patients was low enough to permit escape with comparative ease. Guards also noted that during meals and other times when inmates from different sectors mingled, no distinctive clothing or other markings facilitated distinguishing voluntary inmates from among probationers or prisoners—or vice versa. Recommendations were made to increase surveillance of voluntary patients, to raise the height of the wall in their courtyard, and to restrict their access to recreational spaces.[43]

Kolb was convinced that an enforced stay at Lexington was in the best medical interest of the voluntary patients, but when the possibility and reality of voluntary patients escaping were considered, it was clear that there were some limits on the institution's ability to compel voluntary admissions to stay. To define that limit more clearly, the Public Health Service instituted a test court case in which to make its arguments for forced retention and secure a legal finding on the matter. In correspondence with Treasury Department counsel, Kolb affirmed that voluntary patients entered Lexington having given clear, written consent in the presence of a witness. Moreover, because addiction seemed to spread through association with addicts, these individuals could be considered dangerous to society.[44] The acting attorney general, Stanley Reed, further opined that the 1929 legislation authorizing the creation of the narcotics hospitals reflected Congress's belief that reasonable restraint was a permissible and necessary component of successful treatment.[45] The courts held, however, that patients entering voluntarily could leave the institution at will. Kolb and the classification board found that the mixture of legal constraints on institutional staff and on inmates would become a permanent complicating factor affecting their freedom to make treatment decisions regarding their charges.

Profile of the New Addict in the Lexington Population

Cumming's dedication remarks that addiction spared no social group, rhetorically intended to promote a public view of Lexington as an example of America's charitable treatment of the sick and dependent, belied both the views of Lexington's creators and the demographic reality of opiate addiction by the mid 1930s. The socioeconomic makeup of the Lexington population and its impact on expert thinking about addiction can be discerned by examining a study by Michael Pescor, a PHS psychiatrist on the Lexington staff, published in 1939.[46] Pescor analyzed the patient records for all inmates who entered Lexington between July 1, 1936, and June 30, 1937, a total of 1,036. His objective was to test the validity of the classification scheme Kolb had devised in the 1920s on the basis of his study of about 230 addicts. Did the diagnostic categories assigned to the patients at the end of the initial evaluation period reflect meaningful differences in terms of other traits? Pescor examined such factors as the patients' family backgrounds and socioeconomic circumstances, their medical history, their antisocial record, if any, what drugs they preferred, and what reasons they gave for having become addicted. When he compared these findings to the patients' diagnostic class, he found "significant clusters" within Kolb's categories; thus, he concluded that Kolb's system described a meaningful set of neuropsychiatric diagnostic categories. Pescor's main conclusion was that Kolb's diagnostic category of addiction resulting from psychopathic personality type was a valid new psychiatric diagnosis. He based this conclusion on analysis of a population consisting almost entirely of prisoners and probationers gathered by the criminal justice system. Pescor's study exemplifies Lexington's transitional role in the process whereby addicts came to be seen as so deeply flawed that even lengthy prison sentences could not redeem them.

From his statistics, Pescor developed a hypothetical typical profile for each of Kolb's categories. The first group, normal people accidentally addicted through medical treatment, included just under 4 percent of the sample. The typical member of this category came from a well-adjusted family, with both parents born in the United States, in marginal to comfortable economic circumstances. He had grown up in a rural setting and gave his occupation as farmer. He suffered from some chronic medical condition whose treatment had given rise to an addiction to morphine, and he had entered Lexington as a voluntary patient. His "insight," measured as a recognition that drug use was harmful, was excellent. His prospects on leaving the hospital included returning to a happy marriage and two children, and to a job commensurate with whatever level of disability resulted from his chronic illness. His prognosis was "above average." In discussing his results, Pescor noted that this category "needed no defense."

By contrast, the most numerous category was Kolb's second class, those with a psychopathic diathesis leading to addiction. This group comprised almost 55 percent of the inmates, and it is the most interesting for the present analysis, both because of its size and because of its ambiguous position: the nonmedical origin of the addiction meant these patients were not "normal," but, apart from the addiction itself, they displayed no known psychiatric condition. The issue for Pescor was whether it was valid to consider their personality type as marking a psychopathic flaw that created susceptibility to addiction. Before examining this category in detail, a few salient points regarding the four remaining categories are worth noting. About 6 percent had suffered some form of neurosis before becoming addicted. Although they lacked any history of antisocial behavior, they were troublesome, complaining patients. As children, they had been "studious, shut-in, good boy types" and they had turned to drugs either to relieve pain from physical illness or to assuage their neurotic anxieties. They came largely from the professional or semi-professional classes.

True psychopaths accounted for about 11 percent of the group; they were distinguished from the "psychopathic diathesis" category primarily in that their criminal records predated their drug use and included other offenses besides violations of the Harrison Act. They represented behavior problems in the hospital and typically had to be transferred out of the regular wards into higher security areas in the prison. The inebriates represented the second most numerous group, at about 22 percent of the total, and were generally regarded as well-adjusted, except for their proneness to alcoholism, of which their narcotic addiction was considered a secondary complication. Finally, there was only one case of psychosis predating addiction. Pescor noted that each of these four groups suffered from a condition already well recognized in the psychiatric literature. Since the Type One "normals" were absolved from any hint of psychological predisposition, only the psychopathic diathesis category required justification as a psychiatric diagnosis.

The dissipators or thrill seekers, as Type Two addicts were variously called, differed from the normal accidentals in the following respects in Pescor's profile. They came from economically marginal circumstances rather than from a marginal-to-comfortable range. While the accidentals typically came from rural communities, the Type Two addict came from a deteriorated metropolitan environment, where he worked in "domestic or personal service," usually as a waiter. Hc sometimes engaged in theft to get money to buy drugs. In contrast to the happily married father who prevailed among the accidentals, the dissipator's marriage had typically ended in separation or divorce. There had been no children—perhaps, Pescor opined, because of the deleterious effect of opiates on sexual performance.

The Type Two addict's poor prognosis was based both on the impoverished situation he would return to after release and on his attitudes toward drugs. He had begun drug use either through curiosity or through association with other drug users. He would claim that he had given up drugs forever, but not because he saw them as inherently harmful. On the contrary, he found them beneficial, because they made him feel good. However, he planned to avoid them in order to prevent future incarceration. He would agree superficially with the staff recommendation that after release he plan to live with stable relatives; however, he typically had no job waiting for him, and the risk that he would drift back into association with drug users and then relapse was high.

The Type Two addict's antisocial record consisted solely of drug law violations. His childhood was characterized as normal. In the hospital, he was cooperative and popular among patients and staff. His social adjustment was characterized as satisfactory prior to his addiction, but his tolerance of all forms of vice created a special hazard. In sum, the preaddiction profile of the Type Two addict differed from that of the accidental addict chiefly in terms of socioeconomic background and in attitudes toward drugs. It was only after addiction that the individual's failure to adjust properly became manifest. For this reason, this type was characterized as having a psychopathic diathesis, or predisposition. As Arnold Jaffe has noted, the argument here was circular: the presence of addiction confirmed the preexisting tendency.[47]

Comparison to the psychopathic category is also illuminating. Here, the two types shared socioeconomic circumstances—both came from marginal situations in deteriorated metropolitan areas—but the true psychopath was likely to be an immigrant or first-generation American, while the Type Two thrill seeker was typically a native child of American-born parents. While the thrill seeker had no antisocial record before addiction and worked at a menial service job in an area where vice prevailed, the psychopath already had an arrest and conviction record before the onset of addiction, and his occupation consisted of illegal activities. The thrill seeker appeared to adjust well at Lexington, based on his willingness to work and his congenial relations with fellow inmates and staff. However, his prognosis was poor, as relapse was frequent following return to the dangerous environment. The psychopath, by contrast, frequently had to be transferred out of the regular patient population to a more secure part of the prison.

The Type Two thrill seeker, then, presented a special challenge. His prevalence, in Pescor's view, justified the creation of a separate psychiatric category. Unlike the accidental addict, he was considered problematic; yet his situation did not appear as hopeless as that of the full-blown psychopath. If there was any patient whom Lexington ought to be able to help, it was the Type Two addict. But instead, these individuals typically resumed drug use after discharge, and many then returned to Lexington. The faster-revolving door of short-tem detoxification "cures" seen in the histories of patients at Philadelphia General Hospital in the late 1920s was replaced, for many, by successions of longer prison or probation terms.

Pescor failed to note the irony that the Type Two addicts typically manifested good adjustment to the hospital routine; given the therapeutic focus on adjustment, this pattern should have been grounds for hope. Instead, he saw a probability of relapse in their avowed liking of drug effects and their impending return to their old neighborhoods with no jobs lined up. Moreover, this group's criminal record typically consisted only of drug law violations. Thus, they were diagnosed with a psychiatric condition that was based solely on their addiction, but interpreted as denoting a specific personality type, and they were designated criminals solely because of drug law violations. Pescor correctly noted the risks posed by the typical Type Two's work and residence in areas where the illicit market was centered, but the psychiatric construction put the emphasis entirely on the individual's faults of character and none on the structural nature of the drug market.

The solidification of the idea that addiction betokened underlying character defects occurred at a time when new forms of incarceration were being developed for numerous populations, including the feebleminded and, in a PHS hospital opened in 1932, "defective delinquents." Pescor's study of the first cohort of Lexington inmates provided empirical validation of Kolb's "psychopathic diathesis" category and reified addiction as a psychiatric condition best treated with incarceration. An emerging pattern of recidivism to Lexington further deepened therapeutic pessimism regarding addicts. This pessimism, in turn, helped justify the Lexington mission. A cross between a hospital and a prison was the right place for an addict who was both sick and criminal. The diagnosis lacked a clear etiological foundation but was nevertheless marshaled to justify the incarceration of thousands—at a time when federal policy vigilantly prevented maintenance or research into other forms of treatment, and most physicians welcomed an alternative to treating addicts themselves.[48]

The focus on social adjustment reflected the ideas of reform psychiatrists as they sought to establish professional authority in the context of Progressive Era urbanization, but the trajectory of Kolb's career from diagnostic screener at Ellis Island to medical director of Lexington reflects the difficulties psychiatrists encountered as they tried to resolve the problems they identified. In his research on opiate addiction in the 1920s, Kolb had created a new diagnostic category, the individual whose opiate addiction resulted from a psychopathic personality type. In the late 1920s, he had believed that effective controls on the importation of opiates would protect such individuals from the dangers posed by their psychopathic diathesis. As medical director of an institution intended for the rehabilitation of addicts, where most inmates were sent involuntarily as prisoners or probationers, he, like other psychiatrists of his generation, was forced to confront the limits of the psychopathic diagnosis's utility. Having examined the new American urban environment, psychiatrists had discovered the problematic individual they labeled the psychopath; but by the 1930s, a failure to redeem the psychopath through methods developed in the "psychopathic hospitals" of the Progressive Era had led psychiatrists to resort again to long-term institutionalization as the best course of treatment for these patients, their early identification as "borderline" cases notwithstanding.[49]

Lexington emerged from a coalition of sometimes contradictory ideas, and it reflected the tensions and ambiguities inherent in them. Kolb's psychiatric explanations located the etiology of addiction in individual psychopathology, and inmates were grouped and treatment plans structured around a psychiatric categorizing of addicts. Yet a countervailing belief in the importance of environmental factors was evident in the institutional routine and in the great concern about the environment the patient would return to following release. Thus, the problem etiology was individual psychopathology, but the treatment was environmental. The idea of adjustment resolved this tension between individual etiology and the power of the environment. The patient's illness lay in a failure to adjust properly to his social surroundings. The focus on adjustment allowed recognition of bad social conditions such as deteriorated cities and marginal occupations, but adjustment meant that it was up to the individual, not the social surroundings, to change.

Kolb's humane approach to individual patients notwithstanding, the ideas about addiction he had developed in the mid 1920s provided an appropriate model for an institution whose aim was to change behavioral patterns in inmates deemed socially maladjusted. His characterization of individuals with defective personalities helped bridge the divergent institutional functions of enforcement, incarceration, treatment, and amelioration. In fact, as Kolb had predicted, and as the subsequent history of Lexington proved, a combined hospital-prison became simply a prison. Lexington was thus an institutional expression of therapeutic pessimism at a time when relapse was the expected outcome for most addicts.

As in the asylum for the mentally ill, intractability and recidivism eroded optimism at the PHS Narcotic Hospitals. In the asylum, psychiatrists shifted first to shock treatments (including insulin shock and electroshock therapy) and then to psychosurgery, resorting to powerful and invasive individual therapies rather than relying on the milder impact of routinized institutional life.[50] At Lexington, recidivism reinforced the idea that addicts suffered from a pathological condition that defied the best efforts at cure. Moreover, most of the inmates were federal prisoners serving sentences for violation of drug laws. In the absence of new ideas regarding treatment, the institution became more custodial, and as treatment options outside the PHS hospitals virtually disappeared, addicts were increasingly likely to serve time for their addiction. Within a few years of its opening, Lexington's character as a prison dominated its hospital functions.

Their roles as prisons notwithstanding, the Lexington and Fort Worth Narcotic Hospitals played a crucial role in sustaining Public Health Service involvement in psychiatric training and research. The Division of Mental Hygiene, despite the breadth of scope implied by its name, was originally created as the PHS's Narcotics Division, and through the 1930s, it functioned primarily to manage the PHS Narcotic Hospitals at Lexington and Fort Worth (and to provide medical services in the federal prison system). Lawrence Kolb succeeded Walter Treadway as director of the Division of Mental Hygiene, and he was in turn succeeded by Robert H. Felix, a psychiatrist who had worked at Lexington as part of the Addiction Research Center's team under Himmelsbach. From the late 1930s on, Kolb urged the creation of a national institute, comparable to the National Cancer Institute, dedicated to research on mental illness. The institute was authorized by the 1946 National Mental Health Act, and in 1949, the

Division of Mental Hygiene was replaced by the National Institute of Mental Health, with Robert Felix as its first director.[51] When, following World War II, the PHS began incorporating specialty training throughout its hospital system, the psychiatric residencies at Lexington and Fort Worth were important precedents.[52] Over several decades, a cadre of researchers serving as residents, fellows, or visitors received training in addiction treatment and research at Lexington. Some of these would play important roles in the development of new treatment methods and the reemergence of community-based addiction treatment in the 1960s and 1970s.[53] Examples include Marie Nyswander, co-developer of methadone maintenance treatment, and Jerome Jaffe, who headed the Special Action Office on Drug Abuse Policy in Richard Nixon's administration.

A number of studies in the 1940s and 1950s attempted to determine the success rate of treatment regimens at Lexington and elsewhere, such as California's state hospital for addicts at Spadra. As John O'Donnell has demonstrated, these studies had serious methodological problems—including a tendency to classify even brief episodes of drug use following treatment as "relapse" or treatment failure; but their findings that approximately 80 percent of addicts relapsed sustained the idea that addiction was virtually incurable.[54] These studies bolstered the image of the addict as a psychopathic personality type who belonged in prison. The prevalence of married farmers among those who entered Lexington voluntarily further reinforced the presumed value of the work activities cultivated at Lexington. Eventually, an entrenched pessimism based on therapeutic failures undermined Lexington's therapeutic mission; however, the same failure rate seemed to confirm Kolb's earlier description of addicts as individuals incapable of measuring up to even modest social expectations of productivity and adjustment.

By the 1950s, physicians, scientists, and policy makers agreed that heroin addicts were rarely cured, and little stood in the way of the stiffening sentences for drug trafficking that began with passage of the Boggs Act of 1951. In calling for mandatory minimum sentences for drug trafficking, this law began the dismantling of the Progressive Era reform of individualized sentencing, intended to allow judges to tailor punishment to the criminal rather than the crime.[55] In the decades since, legislators at the state and federal levels have repeatedly passed laws mandating long sentences and reducing judicial discretion for drug offenses.

Kolb's original characterization of opiate addicts as borderline adjustment cases implied hope of reclamation. Instead, both their psychiatric description—bolstered by Pescor's findings—and the addict's criminal status turned liminality into a chasm separating addicts from mainstream American society. On the far side of that chasm, addicts created a social world of their own, which many embraced, desperately or defiantly, as central to their lives.

7

The Addict in the Social Body

The 1950s marked the apogee of American scientific medicine's power and the nadir of status for opiate addicts. Physicians had become paradigmatic of the cultural authority of the professions, while, in both professional and popular venues, "drug addict" was shorthand for profound and unreclaimable deviance. In his magisterial work *The Social System,* the functionalist sociologist Talcott Parsons analyzed medicine to exemplify a disinterested, socially useful profession.[1] The same year that *The Social System* appeared, 1951, Congress passed the Boggs Act, which introduced mandatory minimum sentences to ensure that drug users and sellers would serve hard time.

Also in 1951, another sociologist, Howard Becker, finished his graduate studies at the University of Chicago.[2] As a jazz pianist and graduate student at an elite university, Becker straddled the square and deviant worlds. He had written a master's thesis in which he studied the careers of jazz musicians in much the same ways that his advisor, Everett Hughes, had studied mainstream careers. After completing a dissertation on schoolteachers under Hughes's direction, Becker returned to studying the jazz world, where his intimate familiarity and sense of belonging gave him entrée to the normal workings of a cultural group that was typically exoticized by mainstream observers. Becker had met Alfred Lindesmith, whose University of Chicago dissertation on heroin users, based on research conducted in the 1930s, was published in 1947 under the title *Opiate Addiction.*[3] Becker was impressed primarily with Lindesmith's method, analytic induction, in which theory was relentlessly subordinated to the facts of observation. Hypotheses emerging from fieldwork must fit every observed case or be modified or ruthlessly discarded. Secondarily, Becker saw grounds for an interesting comparison between opiate use, as described by Lindesmith, and marijuana use, which Becker witnessed as a jazz pianist. On the one hand, according to prevailing thought, marijuana was not addictive; on the other, its use involved a socially mediated learning process, as Lindesmith had stressed for heroin.

Becker's marijuana studies were published in the *American Journal of Sociology* and *Human Organization* in 1953. They later appeared, along with two articles about jazz musicians growing from his master's thesis, as chapters in *Outsiders: Studies in the Sociology of Deviance*, in 1963.[4] Becker's work was pivotal in several respects. Conceptually, he broke ground in analyzing a drug-using subculture as a normally functioning social group with its own mores—a group that in part defined itself in opposition to mainstream, conventional culture. Temporally, he drew on the Chicago heritage of urban studies of deviance and, in turn, inspired a later generation of urban ethnographers who studied the worlds of drug use in the 1970s and beyond.

Early sociological studies of drug-using groups contrasted with the psychiatric and pharmacological studies of drug addiction discussed in earlier chapters. The main work on addiction produced in the latter disciplines was consistent with or actively supportive of the federal drug policy of prohibition. In contrast, early sociological studies of heroin addicts produced analyses of drug-using behavior that did not rest on assumptions that addiction was inherently sick or criminal behavior. Sociologists sought to understand the meanings of drug use for those who engaged in it, and they explored the dynamic relationship between drug-using groups and the larger mainstream culture. The result was a body of work that supported a critique of criminal justice management of addiction and the pathologizing of drug use.

Bingham Dai and Alfred Lindesmith at the University of Chicago were the first sociologists to study opiate addiction as social behavior, and the Chicago School of sociology influenced the cohort of urban ethnographers who, beginning about 1970, developed a substantial body of work on patterns of psychoactive drug use in America. To recognize that drug addiction was a profoundly stigmatized condition was not necessarily to note the fact uncritically. In *Outsiders*, Howard Becker repeatedly paired drug addicts and homosexuals as profoundly stigmatized groups that had several features in common.[5] For example, both groups had developed self-justifying ideologies.[6] He stressed that addiction and homosexuality were both conditions that could be hidden, and that exposure could result in serious ostracism, such as loss of one's job.[7]

The ability of heroin addicts and homosexuals to pass as "normals" accounts for the potency of their negative symbolism in the 1950s. On the one hand, the two conditions were constructed as so deviant as to seem utterly incompatible with conventional roles; on the other hand, they could be hidden, so that one might encounter an addict or homosexual and fail to recognize him or her. The parallel to communism at a time when Americans were being alerted to suspect co-workers in the State Department, in labor unions, and in Hollywood studios was not merely implicit; Narcotics Commissioner Harry Anslinger argued forcefully in congressional hearings and the press that the Chinese communists were smuggling heroin into the United States in order to soften up the population in preparation for a takeover.[8]

Concerns about vice had shifted since the Progressive Era, when prostitution had been the focal point of reformers' energies. Then, fears that prostitutes would spread syphilis and gonorrhea to wives and infants fueled the move to suppress prostitution. Progressive Era social hygienists rightly understood the epidemiological reality that sexually transmitted diseases would not remain confined to those whose sexuality was considered deviant, however much some of them cloaked this understanding in eugenic theory and class prejudice. By the 1950s, drug addiction and homosexuality had replaced prostitution as symbolically charged forms of deviance. Concerns about prostitution abated in part because syphilis and gonorrhea no longer presented the threat they once had. The immediate post–World War II period was the time when antibiotics first became widely available to civilian populations (and the advent of antibiotics was a potent marker of the triumph of scientific medicine over age-old human afflictions). The prospect of controlling syphilis and gonorrhea became realistic. At federal and local levels, public health departments mounted programs to identify cases of these sexually transmitted diseases and cure them with the new wonder drugs. Moralism continued to influence public policy regarding sexually transmitted diseases, but it did so within medical and public health circles. The issue no longer had the broad social resonance it had had earlier in the century.[9]

Vice in Chicago Revisited

In 1933, when Walter C. Reckless, a sociologist trained at the University of Chicago, published *Vice in Chicago,* he could still assume that "vice" would connote prostitution.[10] In it, he examined the state of prostitution in Chicago some twenty years after the Chicago vice commission's investigations had culminated in the publication of its report *The Social Evil in Chicago* in 1911.[11] Like its counterparts in other American cities, the Chicago vice commission had recommended legal suppression of prostitution, which was enacted the following year. And, as in other cities, the reformers' nominal triumph ushered in a push and pull, over the course of several mayoral administrations, between proponents of a segregated and tolerated vice district and those committed to suppression. Using the "ecological approach" to the study of urban problems pioneered by the University of Chicago's Department of Sociology, Reckless described the situation as he observed it in the late 1920s and early 1930s.[12]

The suppression policy had not succeeded in eliminating prostitution in Chicago, but it did exert profound changes on how this form of commercial vice was organized and marketed. These changes had occurred in response to shifting political winds; for example, in the 1920s, as the mayor's office switched back and forth between the political machine and the reformers' party, regulation of vice switched with it. In 1923, the reformer William E. Dever defeated the machine politician William Hale Thompson, and Dever's chief of police, Morgan A. Collins, clamped down heavily on vice in the city. One result was the rise of the suburban roadhouse, as proprietors moved their bars and restaurants to where high-stakes rollers gambled and prostitutes trawled beyond city limits. When Thompson was reelected in 1927, parts of the city became "wide open" again.[13] In both cases, the purveying of vice was integrally connected to local and neighborhood politics. Vice entrepreneurs and machine politicians were more obviously linked, but individual police officers took bribes to look the other way and landlords used political clout to shelter illicit activities on their properties even under reform administrations.

The overall trend, Reckless found, was dispersal of a formerly concentrated industry. By 1932, instead of being located in "the old Twenty-second Street district," prostitution had shifted outward to many parts of the city—but had chiefly "invaded neighborhoods of declining respectability."[14] It had also become less visible for various reasons. Brothels had to adopt discreet appearances that belied the business conducted within, and dummy doors and passwords created a protective shield. These stratagems, which the Philadelphia messenger boys quoted in Chapter 1 described as responses to the crackdown on prostitution there, had become standard features of the marketing of vice.

The pattern of dispersal meant that no single neighborhood was dominated by brothels. As the larger houses of prostitution were broken up, new patterns emerged, including women who worked independently as call girls. The prostitute herself had also become less visible. As women had shortened their skirts, cut their hair, and begun wearing makeup, the prostitute was less easy to distinguish from smart young women who worked as stenographers or shop clerks. As Reckless implied, this meant that some women could work discreetly as prostitutes while maintaining other social roles and could make the transition from prostitution to a legitimate way of life relatively easily. But as Ruth Rosen argues, driving prostitution underground broke up the solidarity possible during the days of tolerated vice districts and left many women under the management of pimps and crime syndicates.[15]

Reckless went beyond noting locations of houses of prostitution, as the earlier vice commissions had typically done, to correlating their prevalence to other urban features. He found that neighborhoods where prostitutes tended to live typically had several traits in common, including disproportionately few children compared to the number of adults, low rates of home ownership, and, contrary to popular opinion, higher rates of American-born than immigrant residents. Overall, he concluded that prostitution found a natural home in neighborhoods undergoing disorganization as they declined from prosperity to shabbiness. Among other factors, the process created empty buildings, which landlords would willingly rent on a short-term basis to shady enterprises. This ready availability of cheap real estate facilitated moving from one location to another in response to raids and arrests, and this pattern, in turn, maintained the premium on knowing the people and practices necessary to find one's way to specific purveyors or to guide other customers there.[16]

Overall, Reckless attributed the current pattern of prostitution to "modern urban trends" rather than to the specific actions of law enforcement or reform efforts. He assumed that understanding the spatial distribution of a phenomenon in the city would lead to understanding of the underlying forces shaping the urban scene. His approach exemplified the methods of urban fieldwork developed at the University of Chicago.

The University of Chicago was home to the nation's first department of sociology, founded in 1892, with Albion Small as chair. Although he came to abandon an early infatuation with biological analogies, Small continued to believe they had been useful in suggesting the view of society as a whole system.[17] Small was succeeded by Robert Park, who oversaw the rise of the influential Chicago School of sociology. Chicago sociologists took advantage of their location in the heart of one of America's biggest cities to study the complexities of urban life systematically and firsthand.[18] By mapping various problematic conditions such as suicide, mental illness, or juvenile delinquency across Chicago's neighborhoods and noting where prevalence tended to cluster, these sociologists hoped to demonstrate causative links between types of social environment and incidence of particular problems. For example, Robert E. Faris and H. Warren Dunham examined patterns of mental illness in Chicago, working with a classification of its various neighborhoods developed from census tracts by R. E. Park and E. W. Burgess in *The City*.[19] Faris and Dunham concluded that schizophrenic illnesses tended to concentrate in the marginal urban neighborhoods peopled by transients or residents of low socioeconomic status. By contrast, they found that depressive disorders were distributed much more evenly across the various socioeconomic sectors of the city.

Across the discipline, many sociologists drew on biological models, partly because of the intellectual fit between biological and social functionalism and partly from a desire to link their soft science as closely as possible to the laboratory-based hard natural sciences. In the late nineteenth century, Darwin's evolutionary theories inspired dynamic views of social change through competition, largely through the influence of Herbert Spencer. In the early twentieth century, American sociologists divided along several fault lines, including the role of quantification in social analysis and acceptance or rejection of behaviorism, but they remained united in the belief that scientific tools, correctly deployed, would enable them to make authoritative pronouncements on social ills. Disturbed by what they saw as disorganizing forces in American society before World War I, and disillusioned by the war itself, many embraced social control perspectives in the postwar period.[20] Opiate addiction as an intellectual problem failed to attract the interest of functionalist sociologists. Defined as deviance, it fitted neatly on the wrong side of their normative constructions and posed no conundrums.

At Chicago, however, various factors conduced to an interest in "gritty urban life."[21] A cohort of reformers and academics spent time at Hull House and went on to careers devoted to the study of urban problems. The work, tutelage, and influence of Robert E. Park and William I. Thomas inspired students through the 1920s and 1930s. Thomas had served on the Chicago vice commission. The Rockefeller philanthropies, which had founded the university, later funded the Social Science Research Council in Chicago. Opened in 1923 and modeled on the National Research Council, this body was intended to fund cross-disciplinary social science research. The University of Chicago was a significant beneficiary of SSRC grants in the 1920s.[22]

Two ideas were implicit in the Chicago School approach to social problems: first, that the environment (both the immediate microenvironment in which one strove to belong and the macroenvironment against which smaller groups might define themselves) did much to shape individual behavior through its influences on social roles and status; and, second, that understanding an individual's dynamic trajectory through the developmental phases of a life course and a particular lived history was more important than assigning a fixed character.

As Walter Reckless's *Vice in Chicago* had examined the aftermath of the move to suppress prostitution, another exemplar of this approach, Paul Cressey's *The Taxi-Dance Hall,*[23] examined the licit side of urban amusement. Although his work was still grounded in Chicago sociologists' mixture of attraction to and censure of the semi-legal and illegitimate activities they studied, Cressey employed methods and posed theories that would be taken up again decades later by urban ethnographers studying drug use through participant observation.[24]

Cressey laid out a typology of dance halls, from highly respectable establishments in which young women worked strictly as dance instructors to the taxi-dance halls where men bought tickets, typically at ten cents apiece, for the privilege of dancing for a minute or two with an attractive young woman adept at dance steps and willing to take on any partner who paid. As Cressey said, "Like the taxi-driver with his cab, she is for public hire and is paid in proportion to the time spent and the services rendered."[25] His work focused on this type of dancing establishment, which had sprung up in the wake of the suppression of prostitution. He and his assistants spent time in them, chatting up customers and dancers and using their direct observations to develop an analysis of their social function.

Cressey also developed a typology of men who frequented taxi-dance halls, and he stressed the diversity of types they represented. They were united in one trait, however: lacking legitimate means of socializing with women in a romantic way, they sought such contact through dances and conversation with the taxi-dancers. One feature that distinguished taxi-dance halls from more respectable dancing establishments was the sensuous contact allowed on the dance floor, and some taxi dancers made assignations with customers for later in the night.

In describing the features that could disqualify a man from normal social relations with women, Cressey uncritically reflected the range of stigmas and prejudices characteristic of the period. These included physical handicaps that ranged from disabilities affecting mobility ("crippled") to disfigurement ("pockmarked") to distinctive characteristics ("unusually short or unusually tall").[26] Men of almost any race were common, with the exception that Negroes were excluded. Segregated dance halls for African Americans, or the special case of black-and-tan cabarets, represented a distinct context in the various vice and entertainment markets. The high prevalence of Filipinos among taxi-dance hall customers was explained on the basis of their race and their status as temporary residents of the United States completing professional or business studies; they were expected to return to the Philippines to marry Filipina women. In some cases, though, Filipino men formed liaisons with dance hall women that led to involvements ending variously in marriage or a rupture leaving at least one partner feeling deserted.[27] Other types included recent European immigrants lured by a more open sensuality than allowed in their native cultures, unhappily married men, bohemian globetrotters or slummers, and men too old to be able to attract young women. Cressey argued that for men such as these, the taxi-dance hall performed a useful social function. "The taxi-dance hall is a welcomed institution which grants such a man social acceptance upon equality with all others," he observed. "He embraces it, attends frequently, and seeks to make the most of his new freedom. But at the same time he is conscious of his social limitations. He aspires to venture farther up the social ladder but is restrained by a feeling of inferiority and a baffling sense of inadequacy."[28]

This language and Cressey's statement that "patrons represent, in the main, the great 'lower middle class'"[29] echo Lawrence Kolb's assessment of the typical addict as a man of modest social standing who sought pleasurable means of rising above his limited prospects. However, Cressey was not interested in opiates or addiction and did not include any discussion of drug use. This parallel suggests the problem that men who strayed beyond conventional norms and did not fit neatly into their prescribed status niches posed for sociologists and psychiatrists alike. Kolb saw this character type as the breeding ground for addiction. For Cressey, these men's social difficulties reflected psychological maladjustment, which manifested in inappropriate ways of trying to meet legitimate needs for social contact.[30]

Working in a dance hall offered some of the same attractions that prostitution did for women in the Progressive era. Successful taxi dancers, whom Cressey consistently calls "girls," could earn up to $40 a week, more than factory or store jobs typically paid. They worked just in the evening hours. They had good reason to buy stylish clothes. Cressey saw taxi dancing as a legitimate alternative to prostitution for many women, one that had become available because of the effort to suppress prostitution and because women were allowed more freedom than they had been twenty years earlier. He noted the ease with which young people could live a dual life, engaging in racy or illicit occupations without severing their ties with family or more respectable connections. This freedom also benefited prostitutes, allowing them greater facility to move from prostitution to a more legitimate role. In 1930, women could look and act without censure in ways that would have made them outcasts from respectability in 1910.

Nevertheless, taxi dancing held occupational risks. Like prostitution, taxi dancing could not last long as a career; typically, by her late twenties, a woman had played out her attractiveness and could not compete with newer, younger dancers. For some women, spending a few years as a taxi dancer represented a lucrative interlude between adolescence and marriage, and, Cressey noted, some married women worked as taxi dancers to supplement the conjugal income. But for many, taxi dancing was the first stage in a series of downward status moves, first through seedier dance joints and then into progressively riskier and less remunerative forms of prostitution. Thus, when she began working at a taxi-dance hall, an attractive and "peppy" young woman who danced well would enjoy quick success as a newcomer.[31] In this phase, more experienced dance hall women would show her the ropes, and she would become adept at navigating this social world. In time, though, she would begin to lose status in this setting. Her failure to maintain her success, like her having come to the dance hall in the first place, reflected a maladjusted and disorganized life in which she had failed to sustain normal conventional relationships. Most had worked at menial jobs, had sexual experience, and had had a marriage or intimate relationship fall apart because of infidelities. In time, her situation at the dance hall deteriorated to the point where she moved on to the next, lower-status setting. Here, she went through the same phases again, enjoying initial success, stabilizing for a time, then developing difficulties and shifting down another rung to a new setting. One marker of her downward path was the race of the men she dated. The taxi-dance hall likely exposed her to Filipinos or other Asians such as Chinese; to move from dancing with a Filipino to spending personal time with him was a sign of the loosening hold of conventional norms on a woman's behavior.

Drug Use as Social Behavior

Both the extensiveness of Cressey's direct observation in the field and the dynamic model of downward movement he posited for taxi dancers anticipated by several decades methods and theories that would characterize urban ethnographies of drug-using groups. This did not occur through a direct line of influence extending from his work to that of the cohort of urban drug ethnographers whose work began in about 1970. The earliest sociological studies of opiate addiction as social behavior did, however, emerge from the Chicago School.

Although Chicago sociologists were the first to conduct formal study of opiate addiction as a social behavior, some of the social aspects of drug use had been implicitly recognized in the assumption that drug use that began with "associates," "friends," or "bad company" was diagnostic of the psychopathic addict. Similarly, by the mid 1930s, as the staff at Lexington understood, reexposure to the social world of prior use almost guaranteed resumed use after treatment or prison.

The psychiatrist Pearce Bailey, who had contributed his expertise on heroin users to the World War I psychiatric screening effort, recognized that they created their own social world, writing in 1916 of "a distinct class of heroin addicts, with a certain amount of freemasonry and cooperation among themselves. . . . necessary to make it easy for users to procure heroin and to safeguard one another in the indulgence of a practise strictly forbidden by law. . . . in this way the habit is not only maintained but spreads rapidly." Bailey also recognized that drug use was learned behavior: "The individual habit is formed through the force of imitation and suggestion, and is rarely the result of continuing a drug which was taken in the first instance to allay pain." Bailey described how experienced users would offer heroin powder for sniffing to their friends at dance halls or in city parks, explaining that it was "wonderfully enjoyable," and how the novices would try the drug a second time even when the first use had been unpleasant.[32]

Sociological study of drug addicts at Chicago began in the 1930s when Bingham Dai, then a graduate student, undertook what he saw as a novel approach to studying addiction.[33] He saw drug use as a social behavior that could only be understood by examining the social context within which it occurred. His work played a transitional role in that he brought a new social perspective to bear on understanding drug-using behavior, but he developed his views primarily on the basis of interviews with individuals, usually in institutional settings, rather than on the basis of direct observation of addicts' social interactions.

Bingham Dai's interest in opiate addiction grew out of two experiences during his early life in China. As a young man, under the influence of an uncle, Dai had worked in Shanghai and Nanking with groups seeking to reduce the incidence of addiction through education and prohibitive legislation. Dai was personally affected when he watched the same uncle become an opium addict himself and die without shaking this affliction.

From these experiences, Dai concluded that moral and legal approaches alone were ineffective tools in combating the problem of opiate addiction.[34] Both his exposure to the problem of addiction in a cultural setting other than the American one and his compassionate motivation based on his concern for his uncle may help explain why Dai sought to examine addiction as social behavior reflecting cultural norms. Following his work in China, Dai came to the United States to pursue studies in sociology at Yale's Institute of Human Relations and at the University of Chicago.[35] There, under the mentorship of the eminent criminologist Edwin Sutherland, he developed a proposal to study addiction as a sociological problem—that is, as a problem of the individual in relation to his or her social environment, and growing out of the individual's sense of his or her social role and capacity for fulfilling it.[36]

In November 1933, Sutherland wrote Lawrence Dunham at the Bureau of Social Hygiene on Dai's behalf and enclosed a copy of Dai's research proposal. Dai wished to examine records of addict cases in Chicago, and Sutherland asked whether Dunham would provide an introduction to the director of the Bureau of Narcotics (Harry Anslinger) so that Dai might request permission to examine arrest and conviction reports.[37] At the time, Dai was a student at the University of Chicago on a Rockefeller fellowship. He would use only records of closed cases and reveal no identifying information. Sutherland recommended Dai and the project highly.

For Dunham, "sociology" meant the kinds of surveys of opiate use in various American cities that the Bureau of Social Hygiene had conducted during the 1920s to help determine national levels of medically needed opiates. He saw no value in the project Dai proposed. In his reply to Sutherland, he suggested the only sensible way to proceed was to establish a committee of experts of the caliber of William Charles White, Walter Treadway, and Allan Gregg, and then follow their guidance. To undertake a study without the cooperation of medical authorities such as these would be to risk the embarrassment of producing meaningless results. Dunham explicitly denied Dai's qualification to carry out useful work on the basis of his poorly conceived proposal.[38]

Dai's proposal could hardly have been better calculated to irritate Dunham, since it opened with a statement of the ineffectiveness of the police approach to the addiction problem.[39] Dunham had been the assistant to the New York City police commissioner before coming to the Bureau of Social Hygiene, and he had shifted the bureau's focus from drugs to criminology. Dunham's rebuff did not slow Dai's progress, however. He gained the introduction he sought from Walter Treadway (chief of the Division of Mental Hygiene in the Public Health Service) and received Anslinger's permission to examine the records in question.[40]

Dai's doctoral study of opiate addicts in Chicago, published in 1937, was the first attempt to understand addictive drug use as a social behavior, one conditioned by the individual's relation to his or her social surroundings. For Dai, this social environment had two aspects. One consisted of social norms, in terms of which the individual developed a sense of his or her social role. The individual's own sense of self-worth was based on his or her assessment of how effectively he or she fulfilled this social role. The other component of the social environment consisted of close associates and the circumstances within which the individual lived, behaved, and made choices. This immediate social environment strongly influenced the individual's behavior.

Dai's methods were twofold. First, he undertook a statistical and ecological, or spatial, survey of addicts in Chicago, similar to those of other Chicago sociologists studying the urban distribution of problems such as mental illness or suicide. Using these neighborhood characterizations, Dai found that addicts clustered in particular areas of the city where rents were low, vacancy rates were high, and residents were often transient. The decline of these central city neighborhoods as residential areas followed the growth of industry in the heart of the city. The resulting social milieu was characterized, for Dai, by a loosening of the normative and affectional ties that promoted social organization and harmony.[41] He concluded that addiction was likely to be most prevalent in an environment "in which individuals live mostly by and for themselves, in which the amount of social control is reduced to the minimum, and in which opportunities for unrestrained dissipation and various forms of personal disorganization abound."[42]

Such data revealed the importance of the social milieu in giving rise to drug addiction. However, they could not explain why some individuals responded to such social environments by becoming addicted to opiates while others in the same setting did not. Dai hoped to address this issue by means of a second research method, in-depth interviews with individual addicts. In this choice, he was undoubtedly influenced by William Thomas's insistence that individual life stories would reveal sources of behavior that statistical approaches could not.[43] Working from a psychoanalytic model, Dai expected that skillfully conducted interviews would reveal the childhood traumas, flawed parental relations, or other formative factors that weakened the personality structure of these individuals and predisposed them to addiction. Dai regretted that comparative interview data representative of the general population were not available, but he did not consider this problem fatal to his project. Descriptive data such as his study had yielded were valuable in their own right, he believed.

Dai and his research associates conducted interviews with subjects at the Cook County Psychopathic Hospital and at shelter houses for men; they also located addicts through the Federal Probation Office and through introduction by other addicts.[44] In these open-ended interviews, the subjects were invited to recount their life stories beginning with early childhood memories. In addition to collecting this relevant personal data, the interview situation was also to act as a sort of social laboratory in which the interviewer would assess the subject's behavior, especially as it indicated the subject's attitude toward the researcher. This behavior was seen as an indicator of the subject's attitude toward social situations in general. Dai cited the psychoanalytic idea of transference in noting the importance of behavioral cues such as gestures and facial expressions noted by the interviewer.[45] Correctly interpreted, the subject's response to the interviewer could be viewed as indicative of "the total personality organization of the addict."[46]

Dai provided interview data, including long quoted passages, for twenty-four addicts. In his interpretative remarks regarding individual cases, he ascribed recourse to drugs to failure to adapt successfully to expected social roles. Gender roles and socioeconomic status both represented potentially problematic arenas in which social expectations might clash with individual urges or behavior. The resulting tensions, in Dai's view, drove individuals to seek the relief from emotional pain provided by opiates. One woman had previously worked as an automobile mechanic and later bought the business where she worked. She expressed the desire to reenter this line of business. Dai concluded that, since auto work was a masculine activity in American culture, such an ambition indicated difficulty in accepting her prescribed social role. Her tendency to boast about the local political influence enjoyed by her family further indicated dissatisfaction with her actual social situation. These tensions were further manifested in her attempt to compete with the interviewer, as she announced her intention to write a book about her experiences.

In other interviews, excessive attachment to a parent, or undue harshness in childhood punishments, explained a later inability to make suitable marriages. For males, the desire to compete successfully with other men was often shown to be in conflict with a desire to keep their friendship.[47] Dai consistently viewed homosexual behavior as a sign of difficulty in fulfilling one's rightful social role. Unrealistic aspirations to rise above one's socioeconomic status were also problematic.

Dai framed drug-using behavior, including that which culminated in addiction, as the result of the interaction between the individual and the social environment. He did so in a manner that implicitly accepted the pathology of individuals' misplaced strivings. He accepted a functionalist framework, saying: "The sociological point of view has as its basis the organic conception of the relation between the individual and society."[48] A healthy society was one in which individuals filled mutually supportive and harmonious social roles in a personally satisfying manner. This view placed the same premium on adjustment that Kolb and the psychiatrists building community psychopathic clinics did. Dai's formulation of social roles particularly recalls Lawrence Kolb's description of maladjusted individuals whose ambitions outran their capabilities and who soothed their sense of personal inadequacy with opiates.

Dai worked within certain methodological constraints that linked his work to the policy and bureaucratic sectors where Kolb's ideas had found expression. Although he denied the validity of medical or enforcement analyses of addiction, the institutional world of the medical and criminal justice establishments framed his study. He constructed his statistical profile of addiction in Chicago on the basis of data gathered by the Bureau of Narcotics on arrested drug law violators and by public and private medical treatment facilities. He located most of his interview subjects through psychiatric clinics and probation boards. The social world of addiction, as he was able to indirectly observe it, was thus seen through a set of institutional filters that screened out cases that had avoided capture or exposure to the authorities.

Nor was Dai uncomfortable in assessing this world from the perspective of the dominant culture. Although he recognized the relativism of cultural norms, he judged deviant behavior by prevailing values. In this framework, the onus of social pathology lay entirely on the woman who wanted to own an auto shop, for example, not on the culture that judged her ambition inappropriate. By framing the interview situation as a sort of social laboratory within which the subject's response to social situations could be assessed, Dai implicitly aligned himself and his fellow researchers with the standards of judgment his subjects also encountered in the clinic and the courtroom.

At the same time, Dai made several important conceptual innovations that helped pave the way for later sociological studies of drug-using behavior that would construct the addict's social world in its own terms. Dai's intention was to convey the social world in which the addict encountered opiates, tried them, and developed a habitual pattern of use. He was convinced that understanding the nature of this social world was essential to elucidating the causes of addiction. Because of his own exposure to addiction in two different cultural settings and his training in anthropology, he recognized that cultural norms were socially constructed and that they varied across societies. Furthermore, he recognized that drug-using behaviors were culturally transmitted; for example, he cited historical studies that traced the origins of opium smoking in China to the introduction of tobacco smoking through the development of Dutch trade routes in the Pacific in the eighteenth century. Thus, he argued, nonmedical use of opiates in China dated from the introduction of a drug-use technology (smoking) associated with the use for pleasure of another drug (tobacco). He denied that addiction was a marker of either individual weakness or racial inferiority, as some argued, and he hoped to show that specific social settings promoted the transmission of certain kinds of drug-using behavior.[49]

Here, as elsewhere, Dai introduced a novel framework for conceptualizing drug use, but he remained constrained to the normative structure framing drug use in the American context. Dai noted the social importance of a shift from medical uses of a drug to use for pleasure. He described the social milieu within which recreational drug use occurred. Such settings were characterized by economic marginality, and a prevalence of businesses catering to entertainment, such as restaurants and theaters, which promoted transience among their employees. Dai cited the presence of the illicit drug market as a component of this milieu. He concluded that taking up drug use in such a social setting was an expectable social behavior. For Lawrence Kolb, the difference between medical use and use for pleasure was precisely what separated the normal, or accidental, addict from the pathological addict; the distinction was a key diagnostic marker. Dai discussed the spread of drug use as a form of cultural transmission; public health officials compared it to the contagion of an infectious disease.

Dai thus created the possibility of describing drug use in terms that did not insist on its pathological character. He argued that a purely medical approach to addiction, like the police approach, was too limited to be effective. These perspectives exemplified the two sides of the vice-disease debate. Dai recognized that these conceptions, although mutually contradictory in one sense, both focused on individual behavior without considering the social matrix for that behavior. A law enforcement response was based on the idea that addiction represented willful misbehavior. A medical explanation placed the cause outside the patient's control but still assumed an etiology arising from some individual defect or disfunction. Both assumptions targeted the individual for amelioration, whether through punishment or treatment. (The fundamental compatibility between these notions is apparent in the construction of the Public Health Service Narcotics Hospitals, which functioned both as prisons and as medical institutions.)

Yet, when forced to offer some explanation of why certain individuals in conducive milieux became addicted, while others did not, Dai relied on a psychoanalytic explanation of individual behavior. He concluded that, while association with addicts was crucial in initiating drug use, it was failures of adjustment to marital, gender, occupational, or other roles that caused the individual to develop the drug habit and to relapse repeatedly following attempts at cure. Although he explicitly denied that addiction reflected some innate weakness in certain defective individuals, Dai accepted the premium on social harmony and adjustment that characterized functionalist sociology and the new public health psychiatry. In attempting to identify what factors caused certain individuals to succumb to addiction while others in the same social setting did not, he argued that addiction resulted from the stresses associated with failing to fulfill a satisfying social role. Harmonious and stable family relationships, a stable residence, and steady work all characterized the normative social role against which Dai measured his subjects.

Dai was in part constrained by ideas and methods central to both the sociological and the psychiatric theoretical bases from which he worked. At the methodological level, Dai was forced to rely on institutional records and on individuals' retrospective accounts to create his picture of the addict's social world. His appeal to Dunham and Treadway had been part of an effort to gain access to Bureau of Narcotics arrest records containing information such as age, race, nationality, occupations, drugs used, and age of first use. Dai also consulted records from the Cook County Psychopathic Hospital, a reformatory for women, and a private treatment facility, the Keeley Institute.[50] He separated from the general statistical profiles both the data on women and the data for patients at the Keeley Institute—the latter on the grounds that these private patients differed significantly in socioeconomic profile from the groups who had been arrested or had been treated at the public psychopathic hospital. The Keeley patients included more professionals and had higher incomes than those who had encountered public institutions as a consequence of their drug use. Furthermore, since the Bureau of Narcotics refused to open the records for physicians and pharmacists (individuals licensed under the Harrison Act to handle and dispense opiatcs who had been convicted either for improper use or improper sales), Dai was forced to omit this group from consideration as well. His analysis was thus confined primarily to a sample in which lower socioeconomic groups and individuals whose addiction had come to public attention were overrepresented. Dai's construction of the social world of addiction was thus based almost exclusively on institutional records about institutionalized subjects.

For medical observers such as Kolb who subscribed to the idea that some latent defect explained why some individuals became ensnared in addiction, the individual's behavior was reduced to the effect of innate psychopathology. Dai, like some observers before him, insisted on a cognitive component to addiction. A person was not truly addicted unless he or she recognized that both the pleasurable subjective effects and the torments of withdrawal were associated with the drug. Thus, an individual unknowingly given continuous morphine for pain and then suffering in ignorance through the flulike syndrome following discontinuation of medication would have no basis for resuming drug use. Dai elicited from his subjects accounts of this moment of recognition when sensations were linked to drug knowledge, and he stressed its importance in establishing the pattern of addiction. He also noted that, on first occasions of use, individuals often did not notice or recognize drug effects. In some cases, repeated use occurred before the individual clearly distinguished the drug's impact on sensation. In others, novices took their cues from knowledgeable users in interpreting drug effects.

For Dai, the cues that experienced users gave neophytes to help interpret both initial drug effects and the meaning of withdrawal were crucial; they constituted evidence of the social nature of the behavior and of the transmissibility of its meanings. In fact, both of these moments—the first appreciation of drug effects after one or more false starts and the misunderstanding of withdrawal symptoms until a seasoned addict explained them—had become stock elements of the histories taken from addicts. Roy B. Richardson elicited such accounts repeatedly in his interviews at Philadelphia General Hospital. However, Dai gave these episodes an importance that Richardson did not.

In stressing these affective and cognitive components of drug-using behavior, Dai offered an early model for conceptualizing the subjective drug experience as meaningful in particular ways to the user. Again, the possibility of describing the experience in neutral rather than pathological terms was present. Such a model also opened the door to understanding the drug user's actions as reasonable given the user's own perceptions of his or her social world.

Dai thus broke some conceptual ground in the development of sociological studies of drug-using behavior. He posited meaningful agency on the part of the drug-using individual, but did so within the same normative structure that framed the psychiatric and criminal justice views that he sought to challenge. Because his work combined the germs of a radically new approach to addiction studies with an idea structure shared by normative social science experts of the period, Dai is a transitional figure in the emergence of an ethnographic approach to the study of drug-using behavior.

In 1931, while Dai was still engaged in his research, Alfred Lindesmith came to the University of Chicago to begin graduate studies in sociology. With Herbert Blumer as his dissertation advisor and with guidance from Edwin Sutherland, he, too, began a study of opiate addicts. Lindesmith completed his dissertation in 1937, the same year that Dai published his dissertation. Lindesmith interviewed some of the same subjects that Dai did.[51] Like Dai, Lindesmith stressed the importance of the individual's linking of subjective drug effects to the cognitive awareness that they were produced by the drug; in doing so, he grounded his understanding of addiction in the affective experiences and rational processes of the drug user.

The similarities in the projects notwithstanding, Lindesmith's work marked the beginning of a critique of federal drug policy grounded in disciplinary research. After completing his studies at Chicago, Bingham Dai went on to a career as a psychotherapist, while Lindesmith spent his career as an academic sociologist. Both through his role in educating future sociologists in the theoretical and methodological tenets of symbolic interactionism and through his public stature as a critic of criminal justice approaches to addiction, Lindesmith profoundly influenced sociological study of drug-using groups over the course of the twentieth century.

Based on his study of opiate addicts, Lindesmith concluded that criminal justice sanctions on addictive drug use were cruel and ineffective.[52] An early published account of his research appeared in 1938 in the *American Journal of Sociology*.[53] Here, Lindesmith criticized other accounts of addiction on the basis of faulty method; exceptions could easily be found to disprove them. He rejected the psychiatric theory that addiction arose from a psychopathic personality type. On the one hand, no control studies had been undertaken to determine the prevalence of psychopathy in the population or of psychopaths who were not addicted. On the other, a literature survey revealed ample cases of so-called "normal" individuals struggling unsuccessfully to end an addiction that had begun under medical care. He specifically criticized Lawrence Kolb's work for claiming the importance of psychopathic defects when he found them in only 86 percent of addicts in one study. Besides being excessively deductive and ignoring contrary cases, such theories did little more, in Lindesmith's view, than embody social disapproval of unpopular vices.

Lindesmith further rejected the argument that addicts took opiates as a form of escape or because they were inveterate pleasure seekers. He cited the frequent account by addicts that they needed opiates just to feel normal. Addicts thus were not hedonists but people struggling with powerful cravings. Like Dai, Lindesmith insisted on a cognitive component to this craving. Once an addict experienced the harrowing withdrawal syndrome and learned that a single dose of opiate would dissolve this misery, the compulsion to take opiates to prevent further withdrawal became overpowering. Lindesmith thus linked a cognitive component, knowledge of the significance of the withdrawal syndrome, to a physiological condition, the habituation that had resulted from chronic opiate administration. In doing so, he denied the presence of a particular psychological deficit. In framing the criminalization of opiate use as a function of social prejudice, he mounted a direct critique of the federal policy of prohibition.

However important Lindesmith's commitment to analytical induction was in inspiring other sociologists, his theoretical account of addiction had limitations. By the time Lindesmith was interviewing his subjects, the moment of habit recognition, when an experienced user interpreted the novice's "sickness" and provided an instant cure had long been a stock part of addicts' narratives. Certainly, being told for the first time that one had a habit was a critical moment in assuming the addict identity; but Lindesmith hung far too much definitional weight on this idea. As a result, his view of addiction was somewhat reductionist. Had he done further fieldwork, he might well have probed more deeply into the social contexts that helped shape drug users' identities or used life stories to develop a more nuanced picture of how drug use interacted with other aspects of individuals' lives.

Nevertheless, at the time Lindesmith was writing, the focus on the cognitive understanding of withdrawal created a conceptual opening for examining what addiction meant to addicts; his and Dai's focus on this issue was crucial in charting new directions for examining addiction. Moreover, Lindesmith was as much interested in establishing the validity of analytic induction as he was in advancing any particular theory of addiction—as the method itself dictated. By identifying an issue that applied to every case he examined, he satisfied the rigorous requirements of his method. Comparing his findings to those of other researchers such as Kolb, he found that they could not claim theories that covered 100 percent of their cases.

At the time Lindesmith was conducting his fieldwork interviewing addicts, he accurately perceived that dominant medical opinion about addiction was aligned with federal enforcement policy, and this fact undoubtedly contributed to his rejection of a medical explanation of addiction. Lawrence Kolb, during this time, became medical director of the Lexington Narcotic Hospital, where he still hoped that the regimen for inmates would have a useful rehabilitative effect. Just as Lindesmith was critical of Kolb, Kolb, in turn, thought that Lindesmith completely misunderstood the nature of addiction. After reading Lindesmith's dissertation, he wrote a colleague, "The main theme of his theory is that an addict is not an addict unless he knows it."[54] If addiction were contingent on cognitive awareness of the condition, Kolb argued, then it would be impossible for animals to be addicted; and Kolb himself had conducted animal addiction studies to disprove the autoimmune theory of addiction.[55] Moreover, distinguishing physiological dependence on opiates from a psychological need to use them was crucial to Kolb's distinction between accidental or normal addicts and pathological ones.

In later years, both men shifted positions so as to reduce their grounds for disagreement. In the 1950s and 1960s, a growing group of physicians, lawyers, and others joined Lindesmith in arguing that chronically relapsing addicts should be allowed some form of opiate maintenance; in embracing this position, Lindesmith argued that treating addicts as patients would be more humane and effective than treating them as criminals. Kolb, as we have seen, adopted the same position in these later decades. But in the 1930s, the two men, both prompted by a nonpunitive concern for the welfare of addicts, found themselves on different sides of the policy divide, and a potential alliance was not possible. By the 1930s, too, when Lindesmith was publishing his first work, Charles Terry had already retreated from the policy arena, but Lindesmith cites Terry's advocacy of maintenance with approval in his 1968 book *The Addict and the Law*.[56] Again, a potential alliance between critics of criminalizing addiction did not occur.

Lindesmith's enduring influence in sociology has rested both on his work on opiate addiction and on his influence as a teacher and textbook writer. After completing his dissertation, he never returned to fieldwork. Thus, as a researcher he has always been known primarily for his work on addiction. Lindesmith and Anselm Strauss co-authored the influential, multi-edition textbook *Social Psychology.*[57] Through this vehicle, Lindesmith's ideas on method and on addiction were passed on to successive cohorts of sociology students, beginning in 1949. Lindesmith joined the sociology faculty at Indiana University in 1936, while he was completing the writing of his dissertation, and he spent the rest of his career there.[58]

Lindesmith's reputation as an analyst of addiction rests as well on his long-standing principled critique of federal policy in the face of sustained attempts by the Bureau of Narcotics, and by Harry Anslinger personally, to discredit his work. Anslinger repeatedly engineered the writing of works to oppose positions taken in Lindesmith's writings, and he also attempted to undermine Lindesmith's position as a faculty member at Indiana University.[59]

Biological Models for Social Organization

Within the discipline of sociology, Lindesmith's model of addiction fell within the emerging school of symbolic interactionism and opposed functionalist accounts of addiction. In rejecting the escapist theory of addiction, he set himself against the functionalist view in which addicts, as escapists, were deviants who failed to fulfill useful social roles. The latter view received its most influential expression in Robert Merton's famous account of addicts as double escapists who rejected both socially approved goals and socially approved means of achieving them.[60] For later generations of sociologists, Lindesmith and Howard Becker on the one hand, and Merton on the other, would symbolize these two disparate views of the nature of drug-using behavior.

In the early twentieth century, functionalism was emerging in biology as well as sociology. On one of his trips to the Pondville Hospital in Massachusetts to oversee clinical testing of new opiate analgesics, Clifton Himmelsbach took advantage of his proximity to Harvard to visit the laboratory of Walter Bradford Cannon. Cannon's work on the regulation of body processes and his elaboration of the idea of homeostasis had made him one of the most eminent physiologists in the United States; he was among the scientific notables William Charles White had consulted in the early planning phase of the National Research Council's Committee on Drug Addiction's work.[61] Deeply impressed by what he learned from

Cannon, Himmelsbach came to believe that homeostasis provided the best available model for understanding the physiological phenomenon of addiction.[62]

In the 1930s, endocrinology rather than neuroscience offered a means of uniting behavior and physiology in the same explanatory framework, as hormones were explicated as regulating both physical and emotional states in the body. Homeostasis, which Cannon defined as the "coordinated physiological processes which maintain most of the steady states in the organism," provided a means both of reacting to changed conditions and of returning to a normal, resting state.[63] Whether regulating the pH of the blood or modulating emotion through hormonal negative feedback loops, homeostasis not only conveyed the holism and complexity of the mammalian organism, it also linked physiology with behavior. The complex and internally regulated organism, in turn, provided a metaphor that conveyed the attributes of a functionalist society—a highly complex, highly interactive, civilized social body in which individual actors were expected to absorb and conform to social norms, and in which deviance was almost always unhealthy.

In developing the addictiveness assay, Himmelsbach had exhaustively studied the phenomena associated with opiate withdrawal. He ranged the characteristic symptoms on a severity scale and noted the precise timing with they typically appeared, peaked, and faded away. He correlated the intensity of the withdrawal syndrome with the doses addicts took. Himmelsbach's work had made two familiar aspects of opiate addiction—the development of tolerance with chronic administration, and a physiologically overt withdrawal syndrome—the defining criteria for physical dependence; in turn, these phenomena became the standard against which other drugs were judged physically addictive or not.

This physiologically grounded view of opiate addiction left Himmelsbach well disposed to adopt homeostasis as an explanatory model. The withdrawal symptoms (such as diarrhea) were easily recognizable as the opposites of symptoms caused by the opiates themselves (such as constipation). For Himmelsbach, it appeared that chronic administration of opiates offset the body's internal homeostatic norm for such functions as fluid regulation; when drug use stopped, the measures that had compensated for the drug's presence were still in force, and a period of adjustment was required to reset the homeostatic center. Ironically, Himmelsbach's embrace of homeostasis had some parallel to Lindesmith's idea that addicts repeatedly took opiates to sustain a "normality" artificially induced by tolerance to opiates.

Cannon himself applied the concept of homeostasis metaphorically to the smooth functioning of society. Placing himself in the long line of thinkers who had compared the individual body to the social body, he asked: "May not the new insight into the devices for stabilizing the human organism . . . offer new insight into defects of social organization and into possible modes of dealing with them?" He compared the complexity of modern industrial society, with its "almost unlimited" categories of types of workers, to the complexity of the body.[64] He posited "crude automatic stabilizing processes" in the "body politic":

> A display of conservatism excites a radical revolt and that in turn is followed by a return to conservatism. Loose government and its consequences bring the reformers into power, but their tight reins soon provoke restiveness and the desire for release. . . . Hardly any strong tendency in a nation continues to the stage of disaster; before that extreme is reached corrective forces arise which check the tendency and they commonly prevail to such an excessive degree as themselves to cause a reaction.[65]

The transportation infrastructure that delivered commodities resembled the body's circulatory system, with the water of rivers and canals resembling the blood and lymph. The prime lesson Cannon drew from these similarities was that "*stability is of prime importance.*"[66] Slight disturbances of stability might be warning signals of emerging danger. Although he rejected any reductionist lessons from this comparison, Cannon argued that social stability—maintained through careful management of commerce and industry—left individuals free to pursue their own best fulfillment, just as the "wisdom of the body" in regulating physiological processes freed the individual to think about loftier matters than blood sugar levels.

Sociological Study of Deviance

For Howard Becker and a new group of sociologists at the University of Chicago, drug addicts as research subjects became an important resource for the emerging subfield of deviance studies. For these sociologists, opiate addicts and other users of illicit drugs would be useful in the development of theoretical approaches that challenged the functionalists' organicist model. Deviation from social norms would not be portrayed unproblematically as evidence of social deficiency. Further examination of the worlds of deviant groups would lead to the creation of a new model of society in which groups with authentically different interests engaged in meaningful conflict. When Howard Becker described how jazz musicians and marijuana smokers created their own social worlds, defined in contrast to the dominant, "square" culture, he also stressed the dominant culture's roles in labeling deviance and cutting deviant groups off from the mainstream.[67] This perspective turned upside down the relationship of subculture to dominant culture as it had been portrayed by Bingham Dai. Becker detailed not how just drug users deviated from larger social norms, but how they created their own social world of shared experiences and meanings. Within this world, the behavior of squares took on an anomalous character. Becker took from Lindesmith the idea that drug-using behavior, including appreciation of specific psychoactive effects, was learned behavior that depended on social context. He also adapted the notion of the career from his study of occupational groups and applied it to drug-using behavior.[68] In doing so, he provided a framework for understanding the individual's drug use as a coherent pattern of behaviors linked to identity and life goals. Examining the interactions among social groups and assuming that each group had constructed language, symbols, practices, identities, and social roles consistent with its basic values were important foundations for symbolic interactionism, the theoretical approach Becker, Herbert Blumer, Anselm Strauss, and others developed.

Erving Goffman analyzed stigma as a function of social control and illuminated the situation of individuals striving to move through the social matrix while managing some devalued aspect of their body, their behavior, or their past.[69] For Goffman, drug addiction exemplified a stigmatizing condition, which could be hidden through adroit maneuvering but when revealed disqualified one from a wide range of occupational and personal roles. Dai had suggested that the lying behavior typical of addicts was a function of the addict's social condemnation.[70] Lindesmith argued that addicts lied not because of some innate character defect, but purposively, to achieve specific goals.[71] Goffman characterized such behavior as part of the adaptive strategy of identity management for individuals whose stigmatizing conditions could be hidden; such lies became a tactic in the strategy of passing as a "straight," or normal, individual. The notion of passing provided an analytical tool for relativizing stigma and unearthing the social tensions and conflicts between normative majority behavior and dissenting minority behavior.

In the 1940s and 1950s, sociological framing of drug-using behavior supported a minority consensus of medical and legal professionals that the trend toward increasingly severe penalties for drug use constituted ineffective policy. This view was voiced in a report issued jointly by the American Medical Association and the American Bar Association in 1958, to which Alfred Lindesmith wrote the introduction.[72]

By the late 1960s, when dramatic new patterns of drug use were evident, a new cohort of ethnographic researchers began building on the conceptual tools developed by Becker, Goffman, and others.[73] Ethnographic participant-observer methods were adapted to the study of marginal groups adept at hiding their illicit behaviors; snowball sampling, in which a network of informants is established one contact at a time, was deployed to develop meaningful samples of difficult-to-locate populations. Edward Preble and John J. Casey's classic 1969 article "Taking Care of Business—The Heroin User's Life on the Street" analyzed the street culture of hustling that surrounded heroin acquisition by young minority males in poor New York neighborhoods.[74] They found both that the cultural patterns of drug acquisition were powerfully influenced by changes in the structure of the illicit market and that young minority men developed career roles and status hierarchies through drug buying and selling. In essence, they argued that the conditions put in place by Progressive Era vice reform and the prohibition of opiates had continued to sustain a social context that favored an identity based on street wisdom, quick psychological insight, and good hustling skills. Over the ensuing decades, however, changes in the ethnic composition of urban neighborhoods and shifts in the structures of the illicit market altered that context in various ways, and succeeding generational cohorts of young men, marginalized from mainstream employment opportunities, varied the specifics of their construction of heroin-using and -dealing identities. More recently, Philippe Bourgois's *In Search of Respect: Selling Crack in El Barrio* masterfully combines ethnography, history, and gender and structural analysis to understand a group of young Hispanic male crack dealers in East Harlem.[75]

A profusion of ethnographic studies in the 1970s, many funded by the National Institute on Drug Abuse (created in 1973), focused on specific groups of users as defined by gender, race, class, and neighborhood, as well as on what drugs were used. These moved past the question of whether drug use constituted a pathology (whether medical or criminal) and portrayed it as culturally mediated behavior that, irrespective of the problems it produced, was a desired activity or important coping strategy for individuals and a source of cohesion and shared meaning for social groups.[76] In a study of two distinct PCP-using groups in Miami, Patricia Cleckner showed how class influenced available strategies for managing drug use.[77]

Patrick Biernacki provided the first study of heroin addicts who had given up the drug on their own without undergoing professional treatment or involvement in Twelve Step programs.[78]

Marsha Rosenbaum's study of female heroin users draws directly on Howard Becker's career model to portray the lives of women whose lives became dominated by heroin use. In doing so, she has added important gender analysis to a field that had previously concentrated almost entirely on male addicts. She places heroin use and addiction in the context of the variety of social roles women adopt and are expected to adopt, including those of employee, spouse, and parent. She both describes the neglect of this aspect (for example, noting that in the extensive research on infants born addicted to heroin, the mothers remain peripheral) and argues that women's gender imposes an additional layer of stigma on their addiction.[79]

Methodologically, Rosenbaum draws directly on the work of the "second generation" of Chicago sociologists. Combining an inductive or phenomenological approach with a theoretical commitment to symbolic interactionism, Rosenbaum follows the sequence of women's lives from adolescent experimentation with drugs to full-blown addiction. She argues that as addiction becomes more central to women's lives, and as the strategies necessary to sustain it become more onerous, women become less able both to sustain their conventional roles as workers, wives, or mothers and to maintain their addictions. A progression of such crises creates a negative spiral of narrowing options. Over time, addiction becomes increasingly central to their lives, and the path to a life without drugs becomes increasingly difficult. This progression, in effect, turns the prototypical career path, with its expectation of promotion or advancement, on its head. It also parallels Paul Cressey's account of the downward movement of taxi dancers.

Charles Faupel also analyzes variations in individuals' heroin use over time in conjunction with changes in family relationships, occupation, residence, and other major life issues. As illustrated in the application of Faupel's ideas to Joseph Donatello's heroin addiction in Chapter 4, such analysis shows that an addict's heroin use is not simply driven by physiology but is managed in response to crisis or well-being.[80] These works and scores of others have contributed to a destigmatizing of addiction and a resurgence of treatment resources for drug dependence in the 1970s and 1980s. Although criminal sanctions continued to stiffen (despite some rollback in the 1970s), a nonpunitive treatment infrastructure emerged to meet the growing demand for treatment resulting from the expanded drug use in the 1960s and 1970s. Hospitals rushed to add profitable chemical dependency units in the 1980s, and employee assistance programs and some health insurers covered the cost of treatment and eased the individual's reintegration into family and occupational roles.

For those with the jobs, insurance plans, or financial resources that provide access to such care, addiction can now be managed much like other chronic illnesses according to the social bargain Talcott Parsons outlines in his formulation of the sick role: the addict can take time from normal responsibilities to seek treatment, with the expectation of resuming family and occupational roles (and, incidentally, avoiding incarceration) as long as he or she conscientiously works to get better.[81] Obviously, things rarely work this easily; for example, in spite of sincere efforts at treatment, many addicts alienate close contacts through repeated relapse. Nevertheless, a less stigmatized social role has been available since about the late 1970s for many addicts seeking treatment.[82]

The shelter of a Parsonian sick role was conspicuously absent for addicts in the classic era of narcotic control, however. Lawrence Kolb unwittingly laid the conceptual ground for denying the addict a legitimate sick role when he ascribed relapsing addiction in a person who also engaged in other deviant behaviors to preexisting psychopathology. As discussed in the preceding two chapters, this view facilitated the hardening of federal policy toward addicts through the work of Harry Anslinger and the creation of the two Public Health Service Narcotic Hospitals. At mid-century, when the stigma of being addicted to opiates reached its greatest expression, Talcott Parsons described deviance as a disturbance of social equilibrium that automatically calls into play mechanisms of social control. Like Walter Cannon, Parsons appreciated the challenges facing a society as complex as the United States in the twentieth century, observing: "American society certainly requires an exceptionally high level of affectively neutral and universalistic orientations, both of which are . . . intrinsically difficult of attainment." Parsons exalted both a universalist science and scientific medicine as performing especially crucial social functions. Citing Robert Merton's account of the development of science in the context of early modern rational and ascetic Protestantism, he pointed to education as a distinguishing characteristic of a gentleman: "Knowledge became the most important single mark of generalized superiority."[83] While the prestige of science rested in part on its disinterested search for truth, independent of practical utility, the physician applied science in ways that were both humane and socially useful, as indicated in his discussion of the sick role.

The Social System appeared during the American moment of post–World War II triumph. Not only had science played a crucial role in winning the war, but the recent widespread availability of antibiotics held the promise that medicine might definitively conquer infectious disease. A rapid proliferation of new institutes at the National Institutes of Health exemplified a dramatically expanded federal role in funding biomedical research. In contrast, the recurring finding that 80 percent of addicts typically relapsed had solidified profound pessimism regarding opiate addiction.[84]

Parsons's relatively uncritical acceptance of professional science and medicine in the postwar period, when science and medicine were at the apogee of their power in American society, tacitly endorsed physicians' authority to equate deviance with illness; his psychoanalytically constructed communication system between ego and alter created conceptual room for exploration of norm conformity and norm violation, transference, and identity structure in social interactions. Moreover, his construction of the sick role further granted the physician the competence and authority to sanction release from normal social role obligations in times of verified (and ontological) illness. The patient's side of the bargain consisted in conscientious conformity to prescribed therapeutic regimens to promote a return to health and resumption of usual responsibilities. In such a framework, an addict who would feign illness to gain drugs and feed an addiction was the worst kind of malingerer.

Conclusion

By 1940, all the elements were in place for a configuration of ideas about opiate addiction that remained essentially stable until the enormous demographic changes in drug use that characterized the 1960s. For psychiatrists and pharmacologists, social concerns about opiate addiction in the 1920s and 1930s had created opportunities for disciplinary growth and creation of new knowledge consistent with prevailing theoretical currents in each discipline. In psychiatry, through the work of Lawrence Kolb, a view of addiction consistent with the creation of the new public health psychiatry of adjustment had displaced competing medical views. In pharmacology, formulating a research strategy that linked solving the addiction problem to the development of better medicines had brought resources for creating a research infrastructure to balance the new teaching opportunities resulting from the reform of medical education.

Disciplinary leaders and addiction experts in each discipline saw no inconsistency between their intradisciplinary concerns and the objectives of policy makers in the Public Health Service and the Bureau of Narcotics. Similarly, the addiction research interests of the psychiatrists and pharmacologists in the 1920s and 1930s served the aims of reforming physicians in the American Medical Association leadership. A cluster of related ideas provided a conceptual framework that seemed to confer coherence on the separate efforts of Public Health Service researchers like Lawrence Kolb, pharmacologists seeking improved opiate analgesics, Bureau of Narcotics officials enforcing laws against drug trafficking and possession, and physicians aiming to raise the status of their profession.

For all of these groups, addiction in its current problematic form was associated with the emergence of a complex, modern society. Psychiatrists noted that the increasing complexity of an industrial civilization exacerbated the problems of adjustment. The growing association of opiate addiction with marginal urban neighborhoods, the demimonde of new entertainment and crime venues, seemed to reflect a disparity between social demands and individual capabilities. For pharmacologists, and for policy makers like Lawrence Dunham at the Bureau of Social Hygiene, the development of newer, stronger drugs and the invention of improved drug delivery tools like the hypodermic syringe meant that addiction was a specific hazard unfortunately associated with desirable medical progress. The AMA's long campaign, dating from the 1870s, to gain increased control over the distribution of medications to the public had addressed opiate addiction as a particularly important area of concern. From the late nineteenth century to about 1920, both medical and public opinion agreed that physicians' prescribing practices were chiefly responsible for widespread addiction. By the 1930s, the AMA was able to cite Kolb's theory that the etiology of addiction lay in personality defects, as well as its own activities in restricting the medical uses of opiates, in denying physicians' responsibility for the prevalence of addiction.[1]

A commitment to reducing addiction through controlling the world supply of opiates also linked these interests. The original impetus for passage of the Harrison Narcotic Act had been part of an effort to limit worldwide supplies of opiates to the amounts needed in medical practice. The medical profession approved of keeping opiates out of the hands of the public by means of prescribing laws that gave physicians virtually complete authority to determine who could legally consume opiates. Defining these and other drugs as substances so powerful that only expert medical knowledge could deploy them safely and effectively was entirely consistent with the larger current of medical reform from about 1900 to 1940. The work of the Bureau of Social Hygiene's Committee on Drug Addictions, always consistent with the aim of setting import quotas for opiates, became focused more closely on this activity following Lawrence Dunham's assumption of the committee's directorship.

No unitary explanation of addiction emerged in the period under consideration, and certainly effective control of addiction was not achieved in the form of either definitive cure or foolproof policy. However, Kolb's psychiatric explanation of the addict's personality and the focus of the NRC's Committee on Drug Addiction's pharmacologists on the physiological aspects of tolerance and withdrawal formed a set of ideas that dominated medical and scientific thinking about addiction for several decades. Besides publishing his landmark articles on the psychiatric etiology of addiction in 1925, Kolb worked to disprove physiological theories of addiction that would have challenged his own views. If, for example, Ernest Bishop's and George Pettey's autoimmune theory of addiction were correct, then the process would work identically in all individuals, and Kolb's arguments about specifically susceptible personality types would have collapsed. Kolb's publication (with A. G. DuMez) of an article dismissing the autoimmune theory helped clear the way for wider acceptance of his formulation.

Kolb understood that anyone, given steady administration of opiates for a sufficient period, would develop dependence, but in drawing a distinction between normal individuals accidentally addicted through medication and psychopathic individuals taking drugs for pleasure, he suggested that the universal and physiological aspects of addiction were unimportant in explaining the social problem of addiction. Thus, he had little at stake in ceding this territory to other researchers.

On the physiological front, although little was understood about the mechanism of addiction, there was general consensus that the long-term effects of opiates on various organ systems and tissues were fairly benign. Some investigators, including Oscar Plant at the University of Iowa, whose work was funded by the Bureau of Social Hygiene's Committee on Drug Addictions in the 1920s, hoped to discover something meaningful about addiction by making minute observation of drug effects on various aspects of animal metabolism. However, no findings were sufficiently promising to close off certain avenues of research and highlight others. When Reid Hunt offered a coherent research plan in the area of drug development in hopes of attracting the Bureau of Social Hygiene's proffered funds to the National Research Council, no other research program was compelling enough to compete successfully.

For the team of organic chemists and pharmacologists engaged in the search for a nonaddicting opiate, addiction fell into a well-understood conceptual category: a toxic effect associated with a useful drug. In drug development, both therapeutic and toxic effects were typically assessed in quantitative terms, as researchers sought to maximize beneficial effects and diminish harmful ones. Although drug effects were understood to bear a direct relation to molecular structure, the relationship of structure to activity was not precisely understood. If a toxic effect could not be eliminated, reducing it to acceptable levels was a worthy objective.

In this framework, drug dependence was narrowly defined: the researcher's task was to recognize it when it occurred and to make comparative assessments of its severity. Clifton Himmelsbach and Nathan Eddy's morphine substitution method for test drugs and Himmelsbach's Abstinence Scale for measuring severity of withdrawal syndromes met these needs effectively. These researchers were interested only in the physiological aspects of addiction, because these were what could be unambiguously observed and, to some degree, measured in the laboratory.

Whether a particular psychological makeup made some individuals more prone to addiction than others had no effect on their research interests. The ability of Kolb's psychiatric ideas and the pharmacologists' physiological ones to coexist is exemplified in the location of Himmelsbach's clinical research ward at the Lexington Narcotic Hospital, where Kolb was medical director.

Kolb's psychiatric description of addiction and Himmelsbach and Eddy's characterization of the withdrawal syndrome thus added up to a two-part definition of addiction with complementary parts. A drug was said to be addicting if it produced tolerance with repeated doses and a withdrawal syndrome following abrupt discontinuance. This definition became the criterion for determining whether other classes of drugs were also addictive. Chronic addicts were said to possess innate psychopathic personality traits that made them virtually unreclaimable as healthy and productive citizens.

The conceptual separation of physiological and psychological dependence, an artifact of the different research arenas within which behavioral and physiological addiction were examined, proved long-lasting. Drugs like barbiturates or alcohol, associated with physiologically overt withdrawal syndromes like that linked to opiates, were said to produce physiological dependence. Other drugs associated with compulsive use patterns, like cocaine or marijuana, were described as producing only psychological dependence. Such distinctions reinforced a sense that some drugs were "harder" than others and drew attention away from commonalities of behavior associated with the full range of dependence-producing drugs. They also deflected attention away from different patterns of use of the same drug.

When drug use exploded into new economic, ethnic, and age categories in the 1960s and 1970s, the social underpinnings for these scientific ideas eroded. New social concerns, new settings where addicts were encountered, and a new constellation of disciplinary interests forced reexamination of the meanings of addiction. If the classic era of narcotic control had begun with the forbidding of addiction maintenance and the closing of the municipal narcotic clinics in the 1920s, it ended in the 1960s with the introduction of methadone maintenance as a treatment for heroin addiction.[2]

The idea of methadone maintenance emerged in the context of growing medical dissent from the criminalization of addiction. For it to take hold, the idea that addicts' character flaws would prevent them from improving under a maintenance regimen had to be overturned. This occurred symbolically in Marie Nyswander's embrace of methadone maintenance and her collaboration with Vincent Dole in testing and publicizing the new treatment. During a residency at the Lexington Narcotic Hospital in the 1940s, Nyswander was dismayed by the kinds of treatment she witnessed there and quickly became a campaigner for treating addiction as a disease rather than as a crime. Nevertheless, she absorbed the prevailing belief that addicts' notorious tendency to relapse to drug use following treatment resulted from psychogenic factors. Dole, a metabolic disease specialist, had no such preconceptions; he framed addiction as a problem of physiology altered by continuous drug administration. Methadone, a longer-acting drug than heroin, prevented withdrawal and stabilized the addict's physiology. The longer action also made it easier for people to follow through on educational or occupational trajectories than when they were dependent on heroin, a drug that required repeated administration every four to six hours.[3]

Methadone maintenance was being developed in the early years of an upsurge of drug use by white middle-class youth in the 1960s and 1970s. This pattern led to a demand for an understanding of addiction that did not consign its victims to prison and brand them as having irredeemable character defects. The free clinic movement constituted an early response to the problems posed by young runaways experiencing difficulties with drugs. For example, in 1967, a group including Dr. David E. Smith opened the Haight Ashbury Free Clinic in San Francisco. Smith dropped out of a graduate program in pharmacology to assume clinical directorship of the clinic. From its inception, the clinic was committed to the idea that health care was an inalienable right, which drug use and promiscuity in no way abridged. Inspired in part by the civil rights struggle, this principle also arose from the communitarian and egalitarian ideals of the hippie movement.[4]

By about 1980, a new definition of addiction had been formulated that reflected the changes in patterns of drug use of the 1960s and 1970s and provided professional focus for a burgeoning addiction treatment enterprise into the 1980s. The older essentialist definition, which grounded addiction in unalterable personality characteristics, was replaced by a behavioral one: in a formulation promulgated by researchers and clinicians associated with the Haight Ashbury Free Clinic, addiction was present when drug use was compulsive and out of control and continued in spite of adverse consequences.[5] This more general concept has been applied to nondrug behaviors as well as to drug-using behavior, although distinct pharmacological classes of drugs are recognized as displaying distinct patterns of addiction.

This period saw a rapprochement between professionals like physicians and psychologists, on the one hand, and those involved in treatment methods developed by addicts, on the other. These methods were the Twelve Step approach of Alcoholics Anonymous and the therapeutic community in-patient modality developed by heroin addicts. Recovering alcoholics recognized mainstream medicine's long neglect of addiction. There were all too many stories about doctors and psychiatrists who never asked about drinking or realized that one was an alcoholic. But in the 1970s, when professional groups developed a new interest in addiction, they found in Twelve Step programs, including Narcotics Anonymous, a widespread, decentralized infrastructure in which thousands of individuals participated in a set of ideas and practices that supported restructuring one's life on a basis of abstinence from drug use. The Twelve Step program structure supported consistent involvement that could be incorporated into one's life at varying levels of intensity. These features meant it was well suited for outpatient treatment and in following up detoxification and an initial phase of inpatient treatment. As middle-class demand for addiction treatment grew in the 1970s and 1980s, physicians, psychologists, and a growing cohort of professional addiction counselors established private sector treatment facilities that incorporated Twelve Step meetings or imitated Twelve Step methods.

In the meantime, the policy landscape was profoundly altered after Richard Nixon involved the federal government in the business of funding community-based addiction treatment. When Nixon made law and order a central plank of his 1968 presidential campaign, he saw drug abuse as an important contributor to crime. He was also concerned about the prospect of Vietnam veterans who had become addicted to heroin in Southeast Asia returning to America and adding to the problem of drug-related crime. He appointed Egil Krogh to survey what was known about treating heroin addiction. Krogh identified several hopeful developments. These included Robert Dupont's work with Washington, D.C., jail inmates addicted to heroin and the Illinois Drug Abuse Program, where Jerome Jaffe, a psychiatrist who had worked at the Lexington Narcotic Hospital, was pioneering a multimodality treatment system that included methadone maintenance, therapeutic community facilities, and outpatient counseling.

Concluding that methadone maintenance held great promise as an effective and comparatively inexpensive treatment for heroin addiction, Nixon tapped Jaffe to head a newly created Office of Special Action on Drug Abuse Policy in the White House. Nixon proposed, and Congress passed, legislation to create a National Institute on Drug Abuse (NIDA) in 1972, and the institute was formally created the following year. As planned by Jaffe and as carried out by NIDA, the federal government funded community-based drug treatment centers across the country. In addition to methadone maintenance clinics, NIDA funded long-term residential treatment centers based on the therapeutic community model. In the 1970s, publicly funded methadone maintenance clinics and therapeutic communities were established in cities across the country.[6]

Particularly in the latter setting, as long-time heroin addicts moved out of drug use and confronted the need for conventional occupational roles and social networks that supported abstinence, many became drug counselors. In this way, as envisaged from the first establishment of therapeutic communities, heroin addicts' own knowledge of their experience as addicts became an important source of expertise in shaping and delivering treatment to other addicts entering recovery. The expansion of treatment facilities thus affected the professional structures of drug treatment by creating employment opportunities for addicts in recovery and by normalizing the input of addicts' own knowledge into the development of treatment methods. Here, as in the incorporation of Twelve Step methods into professional treatment venues, the earlier dichotomy between professional expertise and addicts' experience was replaced by systems in which addicts' input was a valued input into new knowledge.

However important and novel Nixon's commitment of federal resources to community-based treatment was, the plan was carried out in a manner that did not challenge the illegality of drug possession or threaten the drug law enforcement establishment. Subsumed under the larger rubric of crime fighting, Nixon's drug treatment initiatives were accompanied by substantial investment in drug interdiction and heightened surveillance of U.S. national borders. Federal funding of treatment began leveling off in 1973, and by mid 1975, enforcement efforts were again funded at higher levels than treatment, both as a share of total government spending on drug policy and in absolute terms.[7]

Elected in 1980 on a platform of reducing the role of federal government, Ronald Reagan further reduced federal funding of drug treatment while supporting intensified enforcement against drug users and drug traffickers. Under his administration, the federal government also reduced its influence over how treatment dollars were spent when it created a block grant system that devolved substantial programmatic responsibility and prerogative to the states. Moves to professionalize the drug-counseling field created crises for former addicts who lacked the educational requirements to meet new standards; some lost their paraprofessional jobs in this process, while trained psychologists and social workers were able to move into the growing field with relative ease (and some former addicts completed the education needed to qualify for more stringently defined jobs). In the same years, Nancy Reagan's Just Say No campaign encouraged a dichotomous separation between abstinence and any use at all of any illicit substance. Similar thinking underlay the "zero tolerance" policy approach of George H. W. Bush's administration.

By the late 1980s, the United States had a two-tier system of responses to individuals' illicit drug use. For the poor, the unemployed, the uneducated, the uninsured, and people of color, drug use was more likely to occur in a setting of concentrated social problems, and users were typically subject to arrest and incarceration. For the middle and upper classes, for those with jobs, education, and health coverage, legal sanctions were less common, and numerous resources were available to support individuals seeking to end destructive drug use. These might include an employer willing to hold a job while the individual completed treatment and health insurance or personal assets to cover the cost of treatment.

The past few decades have also seen broad mainstream acceptance of a loosely defined perception of addiction as the result of difficulty in modulating consumption in an affluent technological society flooded with consumer goods and advertising appeals. Media images of workaholics, chocaholics, and sexaholics parallel the clinical definition of addiction's applicability to almost any substance or activity that people partake of to excess.

Nicotine's current status as an addictive drug in the foreground of national consciousness brings out a number of similarities and contrasts with earlier formulations of opiate addiction. As delivered by tobacco smoking, nicotine has several characteristics that distinguish it from other dependence-inducing psychoactive drugs. First, its use does not provoke socially disruptive behavior. Addictive patterns of use are maintenance patterns: the individual typically maintains a fairly constant blood level by smoking as often as necessary; habits take on fairly stable forms, measurable in numbers of cigarettes or packs a day (in this respect, it resembles opiate addiction). Nicotine does not promote garrulity, violence, or stuporous languor, nor does it interfere with motor function. As a widely consumed licit drug, it is available in socially acceptable locales at reasonable cost (although spaces where one may smoke are increasingly carved out of the surrounding space rather than the other way around). Nicotine addiction in itself is not disruptive of lifestyle or of social order. One could argue, in fact, that its ability to make people feel good in non-disruptive ways is socially useful. Finally, however hard nicotine may be to give up, its withdrawal is not physically debilitating, like withdrawal from opiates, and it does not provoke profound depression, like withdrawal from cocaine. Its most visible aspect is a sniping irritability; this may fray relationships, but such damage is also frequently temporary and reparable. In spite of its mild impact on behavior, however, smoking remains extremely difficult for many people to give up.

Progressive Era observers recognized that smoking tobacco induced dependence, and an anti-smoking campaign formed part of the panoply of Progressive Era reform initiatives.[8] But tobacco failed to ignite the kind of fervor that led to the national prohibitions first of opiates and cocaine and then, for a time, of alcohol. Only since the health costs associated with smoking have been identified as unacceptably high have reformers successfully rolled back broad acceptance of smoking. Again, nicotine is distinctive among psychoactive drugs in that its health consequences are in the distant future, as compared, for example, with alcohol and opiates, which are likely to provoke acute negative effects far earlier in a user's career. Such acute episodes include accidental injury resulting from intoxication and local injuries related to drug administration, such as abscesses at injection sites or perforated septa from snorting, as well as the greater discomforts and, with some drugs, serious health consequences associated with withdrawal.

Thus, it is a future-oriented kind of person who is likely to make the individual choice to give up a profoundly pleasurable habit, smoking, as a hedge against the likelihood of debilitating disease in the later stages of life. In the United States, it is primarily the middle class that has quit smoking, while pushing for policies to prevent unintended exposure to others' smoke.[9] And historically, from the emergence of the bourgeoisie in the seventeenth century to the present day, key characteristics of the middle class have included future-orientation and an affinity for mild stimulant drugs (including, besides tobacco, coffee, tea, and chocolate).[10]

A new disciplinary interest in addiction has arisen among decision theorists, including economists, philosophers, and psychologists, who have introduced new theoretical models in an attempt to understand addiction and to provide means of controlling its associated behaviors.[11] Addiction is of interest in this context because it poses a conundrum to a rational choice model. It presents a fruitful problem for exploring the role of nonrational factors in choice making: how are emotions, drives, cues, or physiological states weighted in an individual's choice-ranking process? These concerns link nicely with work in neuroscience, which has offered models of neurotransmission to explain addiction in ways that remain consistent with a homeostatic metaphor, but that are more complex than the negative feedback loops of hormone regulation. Moreover, with its greater focus on individual behavior than on social relationships, neuroscience mirrors the decision theorist's interest in the processes of individual choice.

For the neuroscientist, the problem of addiction contains the ingredients of successful scientific infrastructure building: the promise of social utility combined with genuine scientific excitement. The National Institute on Drug Abuse funds neuroscientists' work in hope of identifying useful pharmacotherapies for addiction; one research priority is to identify a cocaine maintenance drug to function analogously to the methadone used in maintenance therapy for heroin addicts. For scientists, though, the rewards are not simply those of developing new treatments; studying brain function through such highly technical methods as functional magnetic resonance imaging offers to unlock fundamental mysteries of human nature. The problems of addiction remain central, not only because addictive drugs function in the brain, but also because agonist and antagonist compounds (which trigger or block neurotransmission events) have proved invaluable as probes of neural function. The discovery of endorphins and the elucidation of their role as neurotransmitters, key events in neuroscience, depended in part on the availability of the vast array of opioid compounds developed in the decades-long search for a nonaddicting opiate analgesic, a project originally conceived of as a possible solution to the problem of opiate addiction.[12]

Although physiological models offer reductionist explanations of behavior, some researchers have recently stressed the plasticity of the nervous system, countering more deterministic models of behavior as driven, for example, by genetically determined neurotransmitter levels. Thus, science does not settle the question of volition; rather, it leaves it open for exploration by philosophers, while providing potentially useful models for grounding explanations of behavior. This fact helps account for the ability of neuroscientists and decision theorists to collaborate fruitfully. Neuroscience, like decision theory, however, focuses primarily on the individual as agent, on the individual's weighting of incentives and deterrents, on individual acts of choice or behavior as the unit of interest. This focus underplays the role of social factors (although it does not necessarily exclude them, because, as Talcott Parsons shows, the individual can assign weight to the expected reactions of others to the choices made: my spouse will leave me if I don't stop using drugs; it will be embarrassing if I drink enough to seem out of control). But, even if only to narrow the analytical focus, the more interactive aspects of individuals and norms, the processes whereby norms are constructed, and the kinds of power relationships or cultural values such norms might embody are generally left out of the picture. For historians of knowledge production, these issues take center stage, both for analysis of how they encode power relationships and reveal actors' interests, and for how they shape many kinds of human endeavor, including scientific research.

Drug Prohibition and Public Health

In the 1890s, disturbing trends in the use of dependence-producing drugs and prevalence of a sexually transmitted disease alarmed Americans concerned about social mores and public health. In the closing years of the twentieth century, the transmission of HIV (and, increasingly, of hepatitis C) by shared syringes again entwined stigmatized drug use with an infectious disease stigmatized through its association with sexual deviance. As it confronted the AIDS epidemic, the United States was in the ironic position of having drug laws that impeded work in one of the most basic functions of public health: preventing the spread of infectious disease. Drug paraphernalia laws passed by many states in the late 1970s and 1980s typically included syringes in the list of items whose possession was criminal if the intention was to use them to deliver illicit drugs to the body. These laws were cited as reasons why syringe-exchange programs could not be implemented to reduce the transmission of HIV. Congress passed legislation prohibiting the executive branch from funding syringe exchange unless research demonstrated that it was effective in curbing the spread of HIV, and that it would not lead to increased drug use. Although Health and Human Services Secretary Donna Shalala acknowledged, on April 4, 1998, that these conditions had been met, the Clinton administration nevertheless declined to fund syringe exchange. This situation is a direct legacy of a century's worth of drug policy, born in the Progressive Era.

In the late nineteenth century, reformers framed alcoholism, syphilis, and tuberculosis as a related set of problems endemic to overcrowded and vice-ridden urban working-class neighborhoods. Prostitution was targeted as central to the nexus between illicit sex, syphilis and gonorrhea, and drug use sites such as theaters and saloons. Increasingly, in large North

American cities, cocaine and opiate use became prominent features of this picture of socially corrosive vice. The Bureau of Social Hygiene's creation of a Committee on Drug Addictions as an outgrowth of its work on prostitution exemplifies the perceived links between these issues.

Over the past century, treatment of syphilis and of opiate use, initially linked in reformers' minds, took separate trajectories. After a brief period of optimism and new research initiatives, treatment of opiate addiction became bogged down in pessimism. Syphilis, conversely, was repeatedly reframed in light of advancing biomedical knowledge and practice in the wake of the bacteriological revolution. The discovery in 1905 of the spirochete that causes syphilis and development the following year of the diagnostic Wassermann test brought the disease firmly into the bacteriological purview. When Paul Ehrlich announced the development of the first "magic bullet" to emerge from coordinated chemical and pharmacological research, the drug was a treatment for syphilis. Although Salvarsan was not the safe and definitive cure implied by the rhetoric of the magic bullet, it became powerfully symbolic of the possibilities of new medicines able to cure infectious diseases. In the years after 1905, autopsies confirmed that many of the patients in asylums for the insane had had syphilis, and the syndrome labeled general paresis was definitively identified as a set of complications of tertiary syphilis. These developments, coupled with concerns that syphilis (and gonorrhea) threatened productivity and fertility, helped fuel the Progressive Era social hygiene campaign, the World War I Commission on Training Camp Activities whose mission was to prevent syphilis among American soldiers, and Thomas Parran's anti-syphilis campaigns during his tenures as chief of the PHS Venereal Disease Division, as New York State Health Commissioner, and as surgeon general under Franklin Roosevelt. Sexual mores and the stigma associated with a sexually transmitted disease continually hampered public health efforts at syphilis prevention; in 1934, for example, Parran cancelled a national radio broadcast on venereal disease at the last minute when CBS broadcasters refused to allow him to mention syphilis or gonorrhea by name.[13] However, discovery of a cure for syphilis created a foundation for further progress.

In 1943, Dr. John F. Mahoney of the Public Health Service began investigating the effect of penicillin on syphilis, and following his team's demonstration of the antibiotic's effectiveness, the PHS's Venereal Disease Research Laboratory coordinated numerous studies of different doses, courses of treatment, and so forth to maximize penicillin's effect against syphilis.[14] Racial and social prejudice slowed deployment of penicillin following the discovery of its effectiveness against syphilis,[15] but a vigorous PHS campaign to identify and treat cases drove a steady downward turn in incidence that was apparent by 1950.[16] As the new wonder drugs symbolized the triumph of scientific medicine, they contributed to two forms of complacency: that we were on a steady march of progress toward eradication of all infectious disease, and that once a given disease was under control, vigilance could be relaxed. Progress was such that some physicians suggested overoptimistically that syphilis might be eliminated altogether.

No such definitive remedy was forthcoming for opiate addiction; instead, the image of the junkie as degraded and irredeemable became more deeply entrenched. By the time that the PHS Narcotic Hospitals opened in 1935 and 1938, virtually no other treatment options existed except for private sanitaria, available only to the affluent, and virtually no research on treatment effectiveness was being carried out.[17] The Federal Bureau of Narcotics, under the leadership of Harry Anslinger, moved swiftly to quash local experimentation with addiction maintenance as a treatment method.[18] Drug addiction continued to be a foreground symbol of deviance requiring the most stringent controls—evidenced in passage of the 1951 Boggs Act and the 1956 Narcotic Control Act, with their mandatory minimum sentences and harsh punishments for drug traffickers.

In the 1950s, scientific expertise was king, and the stigmatizing of opiate addicts in medical, moral, and legal contexts seemed socially healthy. At the turn of the twenty-first century, however, knowledge is understood as situated and contextualized not only in academic settings but also in the policy realm of HIV prevention. A more dynamic understanding of human interaction with novel pathogens has erased illusory hopes of wiping out infectious diseases.[19] Current and former drug users' knowledge of specific drug use practices and of the related cultural contexts is seen as crucial public health expertise in disease prevention.

The advent of AIDS punctured the complacency engendered by antibiotics, reminding us that humans live in an ongoing ecological relationship with potential pathogens. This point has become widely recognized and reinforced through outbreaks of Ebola fever and Hanta virus and has prompted concern about the human relationship to the biological environment. A second point, less well recognized, is that AIDS has triggered renewed appreciation of how changes in social arrangements can affect the incidence and prevalence of infectious disease. This awareness echoes a forgotten lesson from the resurgence of sexually transmitted disease in the 1960s. This resurgence is typically blamed, in morally charged terms, on permissiveness and the sexual revolution. However, in the public mind, the simple power of antibiotics has overshadowed the importance of policies and structures that ensure that those who need antibiotics get them. Syphilis was brought under control in the 1950s, not simply by the discovery of antibiotics, but by the creation of a system of venereal disease clinics and public health outreach. As case rates fell and urgency faded, legislatures reduced their funding levels, and this system was substantially dismantled.[20]

The role of social prejudice against gay men and heroin addicts in delaying and blunting public health efforts with regard to AIDS is by now well recognized. However, the increased vulnerability to disease that stigmatization and marginalization can create has received less attention. Patterns of AIDS prevalence in the 1980s revealed the contours of populations driven underground by Progressive Era vice campaigns. More recently, the disproportionate burden of AIDS deaths among African Americans has offered an analytic window for examining how epidemic disease exposes social fault lines.

The history of poverty and the history of race are tightly connected in the United States. The economic effects of deindustrialization hit African American communities sooner than white ones. Residential segregation, sustained by public policy and private sector practice, contributed to a pattern in which poor African Americans were more likely than poor whites to live in neighborhoods of concentrated poverty. Urban renewal projects of the 1950s and 1960s cut freeway paths through African American communities and demolished African American neighborhoods to build convention centers or shopping malls.[21] Given their damaged infrastructures, such impoverished African American neighborhoods were natural sites for markets in illicit commodities like heroin and cocaine in the mid to late twentieth century. The easy availability of drugs, combined with structural barriers to economic opportunity, encouraged addiction among generations lacking significant career prospects.

The disproportionate concentration of HIV infection, AIDS, and AIDS deaths among African Americans is a direct result of this history. As with other social problems, although whites accounted for most of the prevalence of both heroin addiction and AIDS, they were more likely to live in diverse settings not dominated by social problems; in contrast, African Americans were more likely to live in neighborhoods where such problems were concentrated, and thus more visible. (This is not to say that all HIV infections among African Americans are associated with drug use. That male homosexuality has been associated primarily with relatively affluent white men has meant underrecognition of male-with-male sexual contact among working-class men and among African American men. Both of these factors mean that male-with-male sexual contact among African Americans is underestimated. Nevertheless, injection of heroin and cocaine and sexual practices connected with the use of crack cocaine have been critical factors driving HIV infection rates in African American communities.)

In this and other ways, AIDS has pointed up the public health consequences of our draconian drug laws. Criminalizing addiction created a chasm that many drug users fear to cross to seek medical care. Heroin addicts face an array of potential obstacles to care that other patients do not confront. These can include forced withdrawal or detoxification, care from clinicians poorly schooled in managing heroin addiction in conjunction with other diseases, and hostile or judgmental communications from care givers. Laws against syringe possession created a scarcity of syringes, which encouraged sharing and increased the risk of contracting blood-borne infections.

Syringe-exchange programs combine a practical and effective response to the spread of HIV via infected syringes with a philosophical approach, harm reduction, that challenges this legacy of stigmatizing and marginalizing drug users. For those who favor drug prohibition, or who think of syringe exchange as an aspect of drug policy, the controversy over syringe exchange appears as a clash of two policy imperatives: eradication of drug use versus prevention of infectious disease. In the context of harm reduction, the apparent contradiction dissolves, and saving the lives of drug users, their sexual partners, and the infants born to them becomes the highest priority. Although frequently cloaked as drug policy, syringe exchange is better understood as public health policy. In fact, stripped of the symbolic charge surrounding illicit drug use, syringe exchange is simply good old-fashioned infectious disease control: a pathogen has been identified; its life cycle has been studied, and how it is transmitted from one body to another is understood; and syringe exchange interrupts that transmission by providing drug users with sterile injection equipment.

Harm reduction is a broad umbrella covering diverse and sometimes inconsistent views; for example, harm reductionists divide over exactly what form of regulated market for psychoactive drugs would most successfully minimize harms associated with drug use. Some believe that harm reduction is consistent with laws against possession and use of drugs as long as drug treatment is made widely available; others argue for a combination of dismantling criminal justice management of drug use and establishing regulated markets that promote safety in drug use while maximizing support and resources to help people with problems associated with drug use. However, harm reductionists are united in criticizing the United States's punitive policies toward addicts and the demonization of addicts that accompanies such policies. Similarly, they distinguish themselves from libertarian critics who argue for simple abolishment of all drug laws on grounds of individual liberty and a minimal role for government; harm reductionists recognize the risks and public health consequences associated with many forms of drug use and support active public health measures to minimize those harms.

A key point in harm reduction's pragmatic approach to problems associated with risky behaviors is that incremental change in healthy directions is valued even if not all risky behaviors are abandoned. This point challenges the insistence on abstinence that has characterized much drug treatment philosophy over the course of recent decades; it helped account for early tensions between treatment professionals and advocates of syringe exchange. By being willing to work with active drug users even if they continue using drugs, syringe-exchange programs have reached across the breach to provide drug injectors with something they need and want: a reliable supply of sterile injection equipment. Because of the constant need for such supplies, relationships of trust between participants and programs built over time. Syringe-exchange programs have created a point of enduring contact with injection drug users, including homeless people and others seriously disconnected from services. Programs that started out primarily distributing sterile syringes and collecting used ones have broadened to address a wide range of health concerns and have provided important conduits of referral to drug treatment programs and other health and social services. They have been an arena of fruitful interaction between activists, including drug users, and public health professionals and researchers. Through this venue, drug users have contributed their knowledge of specific drug-use practices to the development of health interventions that realistically address the needs that drug users themselves voice in ways that meet the effectiveness standards established through public health research. As ethnographers interested in drug-using behavior developed methods to study "hard to reach" populations, they sometimes created sustained relationships with communities of drug users. These connections provided a foundation for examining the relationship of specific drug-use behaviors and the risk of disease transmission once this issue became urgent.

Research has borne out the effectiveness of harm-reduction-based public health outreach to populations at risk of health complications from drug use. Researchers have monitored injection-related HIV incidence in cities around the world. In cities where syringe exchange was included in HIV-prevention programs targeted at drug injectors when fewer than 5 percent of them were HIV-positive, HIV seroprevalence rates (the percentage of HIV-positive drug injectors) stayed under 5 percent for as many years as the cities were studied. In contrast, cities where syringe-exchange programs were not initiated experienced catastrophic rises in HIV infections. In Bangkok, under an official policy of denial regarding AIDS, the rate of HIV seroprevalence climbed from under 5 to over 40 percent in a single year. A policy reversal in Thailand supported innovative HIV-prevention work that has been a model for other Southeast Asian countries. In New York City, two-thirds of the approximately 200,000 drug injectors were estimated to be HIV-positive in the early 1990s; after the state of New York legalized and funded syringe-exchange programs (in effect ratifying the actions of activists who had been carrying out underground syringe exchanges for several years), infection rates among drug injectors began to fall. Even when not all syringe sharing was eliminated, the reductions in sharing produced by syringe exchange are enough to hold HIV levels steady.[22]

Viewed against the history of the creation of the American junkie, harm reduction reflects a dynamic view of addicts, in which they act in meaningful ways, according to their own situations and values, rather than being static personality types. This view is consistent with the life stories of opiate addicts in the early twentieth century, as well as with both Charles Faupel's model, in which individuals' levels of drug use rise and fall in response to crises or stabilizing factors in their lives, and Marsha Rosenbaum's account of women faced with ever-narrowing options. The rapid success of syringe-exchange programs in reaching large numbers of injectors and the readiness of drug users to alter their behavior in response to threats to their health have helped undermine the view that addicts, whether as a result of personality or drug effect, heedlessly engage in unalterable rituals such as the sharing of syringes. Just as in the early twentieth century, the American heroin addict continues to be an individual managing a burdensome and stigmatizing condition while struggling to maintain social roles based on occupation, gender, family relationships, and so forth. Now, as then, the impact of laws prohibiting drug use and their harsh enforcement has contributed to addicts' alienation from mainstream society and from connection to critical resources such as health care. The risk of HIV infection represents an added burden; like other risks associated with heroin addiction, it is made more dangerous by punitive policies and social marginalization. Ironically, politicians who oppose needle exchange, like earlier generations of their hard-line counterparts, have proven slower to change their behaviors in the face of public health danger than the American junkies who have embraced needle exchange.

Notes

Abbreviations

ANR	Alfred Newton Richards Papers, Archives of the University of Pennsylvania, Philadelphia
DP	Joseph C. Doane Papers, College of Physicians of Philadelphia, Philadelphia
KP	Lawrence Kolb Papers, National Library of Medicine, Bethesda, Md.
LFS	Lyndon F. Small Papers, National Library of Medicine, Bethesda, Md.
NA	National Archives and Records Administration, Public Health Service Records, Record Group 90, National Archives, Washington, D.C.
NRC	National Academy of Sciences/National Research Council Archives, National Academy of Sciences, Washington, D.C., Division of Medical Sciences, Committee on Drug Addiction Record Group
PCCSOA	Records of the Philadelphia Committee for Clinical Study of Opium Addiction, Library of the College of Physicians of Philadelphia, Philadelphia
RAC	Rockefeller Archive Center, Tarrytown, N.Y.

Introduction

1. George Chauncey, *Gay New York: Gender, Urban Culture, and the Making of the Gay Male World, 1890–1940* (New York: Basic Books, 1994).

2. Norman E. Zinberg, *Drug, Set, and Setting: The Basis for Controlled Intoxicant Use* (New Haven: Yale University Press, 1984); Craig Reinarman and Harry G. Levine, "Crack in Context: America's Latest Demon Drug," in *Crack in Amer-*

ica: Demon Drugs and Social Justice, ed. Craig Reinarman and Harry G. Levine (Berkeley: University of California Press, 1997), 1–17.

3. John P. Morgan and Doreen Kagan, "The Dusting of America: The Image of Phencyclidine (PCP) in the Popular Media," in *PCP: Problems and Prevention,* ed. David E. Smith, Donald Wesson, Millicent Buxton, Richard Seymour, Stephanie Ross, Marsha Bishop and E. Leif Zerkin (San Francisco: Haight Ashbury Publications, 1982), 11–20.

4. David T. Courtwright, "Introduction: The Classic Era of Narcotic Control," in *Addicts Who Survived: An Oral History of Narcotic Use in America, 1923–1965,* ed. id., Herman Joseph, and Don Des Jarlais (Knoxville: University of Tennessee Press, 1989), 1–44.

5. Patricia J. Morningstar and Dale D. Chitwood, "How Women and Men Get Cocaine: Sex-Role Stereotypes and Acquisition Patterns," *Journal of Psychoactive Drugs* 19 (1987): 135–42.

6. Edward M. Brecher, *Licit and Illicit Drugs* (Boston: Little, Brown, 1972); Arnold Trebach, *The Heroin Solution* (New Haven: Yale University Press, 1982); Thomas Szasz, *Ceremonial Chemistry: The Ritual Persecution of Drugs, Addicts, and Pushers* (Garden City, N.Y.: Anchor Press, 1974).

7. David T. Courtwright, *Dark Paradise: Opiate Addiction in America Before 1940* (Cambridge, Mass.: Harvard University Press, 1982).

8. Joseph Spillane, *Cocaine: From Medical Marvel to Modern Menace in the United States, 1884–1920* (Baltimore: Johns Hopkins University Press, 2000).

9. Courtwright, *Dark Paradise.*

10. For an account of two Baltimore men—one African American and one white—who became opiate addicts in the years following World War II, see Jill Jonnes, *Hep-Cats, Narcs, and Pipe Dreams: A History of America's Romance with Illegal Drugs* (New York: Scribner, 1996), 240–51.

11. Harvey B. Milkman and Stanley G. Sunderwirth, *Craving for Ecstasy: The Consciousness and Chemistry of Escape* (Lexington, Mass.: Lexington Books, 1987).

12. Richard B. Seymour and David E. Smith, *The Haight Ashbury Free Medical Clinics: Still Free after All These Years, 1967–1987* (San Francisco: Partisan Press, 1986).

13. Caroline Jean Acker, "Stigma or Legitimation? A Historical Examination of the Social Potentials of Addiction Disease Models," *Journal of Psychoactive Drugs* 25, no. 3 (1993): 193–205.

Chapter 1. Heroin Addiction and Urban Vice Reform

1. PCCSOA, vol. 4, case 26–44. Patients' names have been changed.

2. David F. Musto, *The American Disease: Origins of Narcotic Control* (New Haven: Yale University Press, 1973), 30–35.

3. Pearce Bailey, "The Heroin Habit," *New Republic* 6 (Apr. 22, 1916): 314–6, reprinted in *Yesterday's Addicts: American Society and Drug Abuse 1865–1920*, ed. H. Wayne Morgan (Norman: University of Oklahoma Press, 1974), 171–76.

4. Musto, *American Disease*, 54–68.

5. William L. White, *Slaying the Dragon: The History of Addiction Treatment and Recovery in America* (Bloomington, Ill.: Chestnut Hill Systems, 1998), 115, 117.

6. Edward M. Brecher, *Licit and Illicit Drugs* (Boston: Little, Brown, 1972); Arnold Trebach, *The Heroin Solution* (New Haven: Yale University Press, 1982); and Thomas Szasz *Ceremonial Chemistry: The Ritual Persecution of Drugs, Addicts, and Pushers* (Garden City, N.Y.: Anchor Press, 1974).

7. David T. Courtwright, *Dark Paradise: Opiate Addiction in America before 1940* (Cambridge, Mass.: Harvard University Press, 1982).

8. On the anti-tobacco movement, see Cassandra Tate, *Cigarette Wars: The Triumph of "The Little White Slaver"* (New York: Oxford University Press, 1999).

9. Ruth Rosen makes a similar argument for the impact of the suppression of prostitution in this period in *The Lost Sisterhood: Prostitution in America, 1900–1918* (Baltimore: Johns Hopkins University Press, 1982).

10. Timothy J. Gilfoyle, *City of Eros: New York City, Prostitution, and the Commercialization of Sex, 1790–1920* (New York: Norton, 1992).

11. PCCSOA, vol. 8, case 26–15.

12. For a study of gang life about a decade later, see Frederic M. Thrasher, *The Gang: A Study of 1,313 Gangs in Chicago* (Chicago: University of Chicago Press, 1927).

13. The results of the Chicago investigation were published as Vice Commission of Chicago, *The Social Evil in Chicago* (Chicago: Gunthorp-Warren, 1911; reprint; Arno Press/New York Times, 1970.) George Kneeland's New York study was published under the title *Commercialized Prostitution in New York* (1913; reprint, Montclair, N.J.: Patterson Smith, 1969).

14. Lawrence M. Friedman, *Crime and Punishment in American History* (New York: Basic Books, 1993), 127, 225, 341, 426.

15. Vice Commission of Philadelphia, *A Report on Existing Conditions with Recommendations to the Honorable Rudolph Blankenburg, Mayor of Philadelphia* (Philadelphia, 1913), reprinted in *The Prostitute and the Social Reformer:*

Commercial Vice in the Progressive Era, ed. Charles Rosenberg and Carroll Rosenberg-Smith (New York: Arno Press, 1974), 36.

16. Rosen, *Lost Sisterhood,* passim.

17. Friedman, *Crime and Punishment,* 329.

18. Vice Commission of Philadelphia, *Report on Existing Conditions,* 5.

19. Ibid., 80.

20. Ruth Rosen draws these conclusions in *Lost Sisterhood.*

21. Vice Commission of Philadelphia, *Report on Existing Conditions,* 78.

22. Ibid., 162.

23. Ibid., 78.

24. Ibid., 83.

25. Ibid., 78, 82.

26. Dorothy Ross, *The Origins of American Social Science* (Cambridge: Cambridge University Press, 1991).

27. Vice Commission of Chicago, *Social Evil in Chicago;* Walter C. Reckless, *Vice in Chicago* (Chicago: University of Chicago Press, 1933; reprint, Montclair, N.J.: Patterson Smith, 1969).

28. Sir William Osler, "Internal Medicine as a Vocation," in *Aequanimitas, with Other Addresses to Medical Students, Nurses and Practitioners of Medicine,* 3d ed. (New York: McGraw-Hill Blakiston Division, 1932), quoted in Allan M. Brandt, *No Magic Bullet: A Social History of Venereal Disease in the United States since 1880* (1985; rev. ed., New York: Oxford University Press, 1987), 10.

29. Brandt, *No Magic Bullet,* 7–51; Elizabeth Lunbeck, *The Psychiatric Persuasion: Knowledge, Gender, and Power in Modern America* (Princeton: Princeton University Press, 1994), 71–72.

30. John Burnham analyzes the whole constellation of vices in this period, and the commercial forces that encouraged them, in *Bad Habits: Drinking, Smoking, Taking Drugs, Gambling, Sexual Misbehavior, and Swearing in American History* (New York: New York University Press, 1993).

31. Lunbeck, *Psychiatric Persuasion,* 126–27.

32. Jack Pressman, *Last Resort: Psychosurgery and the Limits of Medicine* (Cambridge: Cambridge University Press, 1998), 33.

33. Brandt, *No Magic Bullet;* Elizabeth Fee, "Sin Versus Science: Venereal Disease in Twentieth-Century Baltimore," in *AIDS: The Burdens of History,* ed. id. and Daniel M. Fox (Berkeley: University of California Press, 1988), 121–46.

34. Michael Gottlieb and Joel Weisman, "*Pneumocystis* Pneumonia—Los Angeles," *Morbidity and Mortality Weekly Report,* June 5, 1981; Gerald Oppenheimer, "In the Eye of the Storm: The Epidemiological Construction of AIDS," in *AIDS: The Burdens of History,* ed. Elizabeth Fee and Daniel M. Fox (Berkeley: University of California Press, 1988).

35. Thrasher, *Gang.*

36. *The City Wilderness: A Settlement Study by Residents and Associates of the South End House,* ed. Robert A. Woods (Boston: Houghton, Mifflin, 1898), quoted in Thrasher, *Gang,* 265.

37. Marsha Rosenbaum, *Women on Heroin* (New Brunswick, N.J.: Rutgers University Press, 1981).

38. PCCSOA, vol. 7, case 99–01. Case numbers beginning with "99" represent unnumbered cases in the PCCSOA files. I assigned these case numbers myself; volume numbers indicate location of the files. A complete list of case numbers cited in this book is available at the College of Physicians of Philadelphia.

39. Arthur B. Light, M.D., to Captain Francis J. Dunn, Dept. of Public Safety, Philadelphia, Feb. 16, 1929, PCCSOA, vol. 30.

40. PCCSOA, vol. 10, case 26–103.

41. Paul G. Cressey, *The Taxi-Dance Hall: A Sociological Study in Commercialized Recreation and City Life* (Chicago: University of Chicago Press, 1932; reprint, New York: Greenwood Press, 1968); Reckless, *Vice in Chicago.*

42. Caroline J. Acker, "From All-Purpose Anodyne to Marker of Deviance: Physicians' Attitudes Toward Opiates in the U.S., 1890–1940," in *Drugs and Narcotics in History,* ed. Roy Porter and Mikulas Teich (Cambridge: Cambridge University Press, 1995), 114–32.

43. Samuel Hopkins Adams, "Part IV: The Subtle Poisons," *Collier's,* Dec. 2, 1905, 16.

44. James Harvey Young, *The Medical Messiahs: A Social History of Health Quackery in Twentieth-Century America* (Princeton: Princeton University Press, 1967).

45. Musto, *American Disease,* 91–120.

46. William B. McAllister, *Drug Diplomacy in the Twentieth Century: An International History* (New York: Routledge, 2000); Musto, *American Disease,* 24 ff.; Brecher, *Licit and Illicit Drugs,* 48–49.

47. See, e.g., Edward Huntington Williams, *Opiate Addiction: Its Handling and Treatment* (New York: Macmillan, 1922).

48. Musto, *American Disease,* chs. 5–7; Dan Waldorf, M. Orlick, and Craig Reinarman, *Morphine Maintenance: The Shreveport Clinic, 1919–1923* (Washington, D.C.: Drug Abuse Council, 1974).

49. Lawrence Kolb and A. G. DuMez, "Prevalence and Trend of Drug Addiction in the U.S. and Factors Influencing It," *Public Health Reports* 39, no. 21 (May 23, 1924): 1179–1204.

50. Harry M. Marks, *The Progress of Experiment: Science and Therapeutic Reform in the United States, 1900–1990* (Cambridge: Cambridge University Press, 1997).

51. American Medical Association, *Nostrums and Quackery and Pseudo-Medicine,* 3 vols. (Chicago: Press of American Medical Association, 1911–36).

52. Acker, "All-Purpose Anodyne."

53. Caroline Jean Acker, "Stigma or Legitimation? A Historical Examination of the Social Potentials of Addiction Disease Models," *Journal of Psychoactive Drugs* 25, no. 3 (1993): 193–205.

54. Charles E. Terry and Mildred Pellens, *The Opium Problem* (New York: Bureau of Social Hygiene, 1928), 167–311, passim.

55. Glenn Sonnedecker, *Emergence of the Concept of Opiate Addiction* (Madison, Wis.: American Institute for the History of Pharmacy, n.d.), 12–13.

56. Terry and Pellens, *Opium Problem,* 154.

57. C. F. J. Laase, quoted in Terry and Pellens, *Opium Problem,* 155.

58. Terry and Pellens, *Opium Problem,* 155–60.

59. Committee on Narcotic Drugs of the Medical Society of New York, quoted in Terry and Pellens, *Opium Problem,* 161.

60. Bishop's ideas are laid out in numerous sources, including *The Narcotic Drug Problem* (New York: Macmillan, 1920); "Narcotic Addiction—A Systemic Disease Condition," *Journal of the American Medical Association* 60 (1913): 431–34; and "An Analysis of Narcotic Drug Addiction," *New York Medical Journal* 101 (1915): 399–403. Another formulation, independently developed, appears in George E. Pettey, *The Narcotic Drug Diseases and Allied Ailments: Pathology, Pathogenesis, and Treatment* (Philadelphia: F. A. Davis, 1913).

61. Bishop, "Analysis of Narcotic Drug Addiction," 401.

62. Joseph C. Doane, "A Brief History of the Philadelphia General Hospital from 1908 to 1928," in *History of Blockley: A History of the Philadelphia General Hospital from Its Inception, 1731–1928,* comp. John Welsh Croskey, M.D. (Philadelphia: F. A. Davis, 1929).

63. "New Drug Law in Force: Hard for the Fiends," *Philadelphia Press,* Mar. 2, 1915 (clipping in scrapbook), DP, box 3.

64. Ibid.

65. Musto, *American Disease,* 80–90; Charles B. Towns, *Habits That Handicap: The Menace of Opium, Alcohol, and Tobacco, and the Remedy* (New York: Century, 1916).

66. Judith Walzer Leavitt, *Brought to Bed: Child-Bearing in America, 1750–1950* (New York: Oxford University Press, 1986).

67. Joel Braslow, *Mental Ills and Bodily Cures: Psychiatric Treatment in the First Half of the Twentieth Century* (Berkeley: University of California Press, 1997), 36.

68. "Wholesale Cure of Dope Fiends," *Philadelphia North American,* Mar. 28, 1915 (clipping in scrapbook), DP, box 3.

69. Joseph C. Doane, "Drug Inebriety," *Medical Clinics of North America* 3 (1920): 1483.

Chapter 2. The Opportunistic Approach

1. Ruth Rosen, *The Lost Sisterhood: Prostitution in America, 1900–1918* (Baltimore: Johns Hopkins University Press, 1982), 14–15.

2. Allan M. Brandt, *No Magic Bullet: A Social History of Venereal Disease in the United States since 1880* (1985; rev. ed., New York: Oxford University Press, 1987), 38; Ellen Fitzpatrick, *Endless Crusade: Women Social Scientists and Progressive Reform* (New York: Oxford University Press, 1990), 111.

3. Brandt, *No Magic Bullet,* 39.

4. Fitzpatrick, *Endless Crusade,* 110–12.

5. Adele E. Clarke, *Disciplining Reproduction: Modernity, American Life Sciences, and the Problems of Sex* (Berkeley: University of California Press, 1998).

6. Robert E. Kohler, *Partners in Science: Foundations and Natural Scientists, 1900–1945* (Chicago: University of Chicago Press, 1991).

7. On how such a plan worked to the interest of scientists at elite institutions, see David A. Hounshell, "The Evolution of Industrial Research in the United States," in *Engines of Innovation: U.S. Industrial Research at the End of an Era,* ed. Richard S. Rosenbloom and William J. Spencer (Boston: Harvard Business School Press, 1996), 13–85.

8. William B. McAllister, *Drug Diplomacy in the Twentieth Century: An International History* (New York: Routledge, 2000).

9. David T. Courtwright, "Charles Terry, *The Opium Problem,* and American Narcotic Policy," *Journal of Drug Issues* 16, no. 3 (1986): 421–34.

10. "Conferences held at Gedney Farms," RAC, record group 1.1, ser. 200, box 16, folder 183.

11. David F. Musto, *The American Disease: Origins of Narcotic Control* (New Haven: Yale University Press, 1973), 319 n. 12; see also p. 186.

12. Lawrence Kolb and A. G. DuMez, "Prevalence and Trend of Drug Addiction in the U.S. and Factors Influencing It," *Public Health Reports* 39, no. 21 (May 23, 1924): 1179–1204; David T. Courtwright, *Dark Paradise: Opiate Addiction in America Before 1940* (Cambridge, Mass.: Harvard University Press, 1982).

13. R. B. Fosdick to J. D. Rockefeller Jr., Mar. 18, 1921, RAC, record group 2, ser. 7, folder 51.

14. Frederic M. Thrasher approvingly describes Woods's crime prevention work with juveniles during his tenure as police commissioner in *The Gang: A Study of 1,313 Gangs in Chicago* (Chicago: University of Chicago Press, 1927), 529.

15. On Davis's life and career, see Fitzpatrick, *Endless Crusade;* Clarke, *Disciplining Reproduction,* 93–97.

16. "Report of the Sub-Committee . . . Jan. 30, 1924," RAC, Bureau of Social Hygiene record group, ser. 3, subser. 1, box 4, folder 140.

17. McAllister, *Drug Diplomacy,* 23–24.

18. "Committee on Drug Addiction . . . Tentative Budget, 1925," Jan. 1, 1925, RAC, record group 2, Rockefeller Boards ser., Bureau of Social Hygiene subser., box 7, folder 51.

19. "University of Pennsylvania—Professor A. N. Richards" (n.d.), ANR, box 11, folder 32; "Oscar Henry Plant, 1875–1939," *Journal of Pharmacology and Experimental Therapeutics* 67, no. 3 (1939): 365–71.

20. C. E. Terry to W. S. Richardson, July 20, 1925, RAC, record group 2, Rockefeller Boards ser., Bureau of Social Hygiene subser., box 7, folder 51.

21. Alfred Newton Richards, "The Opium Problem," "The Scope of the Committee," etc. (typescripts, n.d.), ANR, box 11, folder 41.

22. Alfred Newton Richards to Charles Terry, July 22, 1925, ANR, box 11, folder 24, and June 9, 1928, ANR, box 11, folder 26.

23. Remarks of Joseph Doane, "Chemical and Clinical Studies of Opium Addiction: Program given before the Philadelphia County Medical Society Wednesday, December 19th, 1928, sponsored by the Philadelphia Committee for the Clinical Study of Opium Addiction" (typescript), ANR, box 11, folder 29.

24. "The First Annual Summary of the Clinical Work on Opium Addiction," RAC, Bureau of Social Hygiene collection, ser. 4, subser. 1, box 1, folder 552.

25. Arthur B. Light et al., *Opium Addiction* (Chicago: American Medical Association, 1929–30).

26. Charles Terry and Mildred Pellens, *The Opium Problem* (New York: Bureau of Social Hygiene, 1928).

27. Ibid., 89–91.

28. Ibid., 90.

29. Ibid., 91.

30. On the impact of Harrison Act enforcement in cutting off the accumulation of knowledge about treating addiction, see William L. White, *Slaying the Dragon: The History of Addiction Treatment and Recovery in America* (Bloomington, Ill.: Chestnut Health Systems, 1998). On federal action to oppose local maintenance initiatives in California, see Jim Baumohl, "Maintaining Orthodoxy: The Depression-Era Struggle over Morphine Maintenance in California," in *Altering American Consciousness: Essays on the History of Alcohol and Drug Use in the United States, 1800–1997,* ed. Sarah W. Tracy and Caroline Jean Acker (Amherst: University of Massachusetts Press, forthcoming).

31. Katharine Bement Davis, "Memorandum," Jan. 26, 1926, RAC, record

group 2, Rockefeller Boards ser., Bureau of Social Hygiene subser., box 7, folder 51.

32. Minutes, CDA meeting, Jan. 30, 1924, RAC, Bureau of Social Hygiene, ser. 3, subser. 1, box 4, folder 140.

33. Brandt, *No Magic Bullet,* 7–23.

34. Terry and Pellens, *Opium Problem,* 91.

35. See Edward Huntington Williams, *Opiate Addiction: Its Handling and Treatment* (New York: Macmillan, 1922). On Williams's efforts in support of maintenance in California, see Baumohl, "Maintaining Orthodoxy."

36. Courtwright, "Charles Terry," 429.

37. "Survey of Scientific Opinion," RAC, Bureau of Social Hygiene collection, record group 4, subser. 1, box 1, folder 54.

38. Minutes, CDA meeting, Dec. 17, 1917, ANR, box 11, folder 29.

39. "Memorandum," L. B. Dunham to A. Woods, Apr. 27, 1926, RAC, Bureau of Social Hygiene collection, ser. 3, subser. 1, box 5, folder 149.

40. K. B. Davis, "A Programme for the Committee," Dec. 18, 1927, RAC, Bureau of Social Hygiene collection, ser. 3, subser. 1, box 4, folder 140.

41. See RAC, Bureau of Social Hygiene collection, ser. 3, subser. 1, box 6, folder 159, for this correspondence (cited in nn. 42–44 below).

42. L. B. Dunham to F. P. Underhill, Mar. 16, 1927.

43. Underhill to Dunham, Mar. 18, 1927.

44. Dunham to Underhill, May 6, 1927.

45. Frank P. Underhill, "Report of Investigation on 'Ear-Marking' of Drugs—1927–1928," RAC, Bureau of Social Hygiene collection, ser. 3, subser. 1, box 6, folder 159.

46. Such concerns were raised by Terry (L. B. Dunham to K. B. Davis, June 2, 1927, RAC, Bureau of Social Hygiene collection, ser. 3, subser. 1, box 4, folder 141) and Lawrence Kolb (Kolb to L. G. Nutt, Aug. 1, 1928, ibid., box 5, folder 149). Kolb also believed that enforcing the ear-marking requirement would prove unfeasible.

47. Dunham to Davis, June 2, 1927, RAC, Bureau of Social Hygiene collection, ser. 3, subser. 1, box 4, folder 141.

48. Charles Terry, "To the Chairman and Members of the Committee on Drug Addictions," Oct. 30, 1927, RAC, Bureau of Social Hygiene collection, ser. 3, subser. 1, box 4, folder 141.

49. Dunham to Vernon Kellogg, Dec. 8, 1928, RAC, Bureau of Social Hygiene collection, ser. 3, subser. 1, box 4, folder 141.

50. Dunham to William Charles White, 12 Jan. 1931, RAC, Bureau of Social Hygiene Collection, ser. 3, subser. 1, box 5, folder 144.

51. Dunham to Colonel Woods, Jan. 22, 1930; Farnham to Dunham, Apr. 2,

1931, both in RAC, Bureau of Social Hygiene collection, ser. 3, subser. 1, box 6, folder 152.

52. Courtwright, "Charles Terry."

Chapter 3. The Technological Fix: The Search for a Nonaddicting Analgesic

An earlier version of this chapter appeared as "Addiction and the Laboratory: The Work of the National Research Council's Committee on Drug Addiction, 1928–1939," *Isis* 86 (1995): 167–93.

1. See Caroline Jean Acker, "From All-Purpose Anodyne to Marker of Deviance: Physicians' Attitudes Towards Opiates in the U.S. from 1890 to 1940," in *Drugs and Narcotics in History,* ed. Roy Porter and Mikulas Teich (Cambridge: Cambridge University Press, 1995), 114–32.

2. Reid Hunt to Victor C. Vaughan, 10 July 1922, NRC, Central File, Division of Medical Sciences: Projects, MED: Projects 1922, Pharmacology Studies; Proposed folder.

3. Timothy Lenoir, *Instituting Science: The Cultural Production of Scientific Disciplines* (Stanford, Calif.: Stanford University Press, 1997)

4. Cf. Charles E. Rosenberg, "Toward an Ecology of Knowledge: On Discipline, Context, and History," in *The Organization of Knowledge in Modern America,* ed. Alexandra Oleson and John Voss (Baltimore: Johns Hopkins University Press, 1979), 440–55.

5. Susan Leigh Star and James R. Griesemer, "Institutional Ecology, 'Translations,' and Boundary Objects: Amateurs and Professionals in Berkeley's Museum of Vertebrate Zoology, 1907–19," *Social Studies of Science* 19 (1989): 387–420.

6. Vaughan to Hunt, July 13 and 28, 1922, both in NRC, MED: Projects 1922, Pharmacology Studies; Proposed folder.

7. Hunt to Vaughan, July 26, 1922; Hunt, "Suggestions for the Promotion of Pharmacology in the United States," enclosed with Hunt to Vaughan, Aug. 2, 1922; all in NRC, MED: Projects 1922; Pharmacology Studies; Proposed folder.

8. Executive Committee, Division of Medical Sciences, motion, Sept. 22, 1922, in NRC, MED: Projects 1922; Pharmacology Studies; Proposed folder.

9. On Kellogg's career, see C. E. McClung, "Vernon Lyman Kellogg," *Biographical Memoirs* (Washington, D.C.: National Academy of Sciences) 20 (1939): 245–58.

10. Dunham to Vernon Kellogg, Nov. 23, 1928, NRC, MED: Com on Drug Addiction 1928–1929, Beginning of Program folder.

11. White to Dunham, Nov. 26, 1928, NRC, MED: Com on Drug Addiction 1928–1929, Beginning of Program folder.

12. White to Division members (telegram), Dec. 10, 1928; replies from members; Kellogg to Brockett (assistant secretary, National Academy of Sciences), Dec. 11, 28; all in NRC, MED: Com on Drug Addiction 1928–1929, Beginning of Program folder. The NRC preferred the singular "Addiction" to the Bureau of Social Hygiene's "Addictions."

13. White to Kellogg, "Memorandum," Jan. 14, 1929, NRC, INSTITUTIONS Assoc Individuals 1928–1930, Bureau of Social Hygiene, Com on Drug Addictions, Termination of Activities folder.

14. On Sollmann's stature and activities, see John Parascandola, *The Development of American Pharmacology: John J. Abel and the Shaping of a Discipline* (Baltimore: Johns Hopkins University Press, 1992), and John P. Swann, *Academic Scientists and the Pharmaceutical Industry: Cooperative Research in Twentieth-Century America* (Baltimore: Johns Hopkins University Press, 1988).

15. Hunt to Voegtlin, Dec. 17, 1928, NRC, MED: Com on Drug Addiction 1928–1929, Beginning of Program folder.

16. John Parascandola, "Charles Holmes Herty and the Effort to Establish an Institute for Drug Research in Post World War I America," in *Chemistry and Modern Society: Historical Essays in Honor of Aaron J. Ihde,* ed. John Parascandola and James C. Whorton (Washington, D.C.: American Chemical Society), 85–103; Victoria A. Harden, *Inventing the NIH: Federal Biomedical Research Policy, 1887–1937* (Baltimore: Johns Hopkins University Press, 1986), 71–91. The committee's report was published as John J. Abel, Carl L. Alsberg, Raymond F. Bacon, F. R. Eldred, Reid Hunt, Treat B. Johnson, Julius Stieglitz, F. O. Taylor, and Charles H. Herty, *The Future Independence and Progress of American Medicine in the Age of Chemistry* (New York: Chemical Foundation, 1921).

17. William Bynum, "Chemical Structure and Pharmacological Action: A Chapter in the History of Nineteenth-Century Molecular Pharmacology," *Bulletin of the History of Medicine* 44 (1970): 518–38; John Parascandola, "The Development of Receptor Theory," in *Discoveries in Pharmacology,* vol. 3: *Pharmacological Methods, Receptors, and Chemotherapy,* ed. Michael J. Parnham and J. Bruinvels, 129–56 (New York: Elsevier, 1986).

18. See Lenoir, *Instituting Science,* ch. 7, "A Magic Bullet: Research for Profit and the Growth of Knowledge in Germany Around 1900"; Georg Meyer-Thurow, "The Industrialization of Invention: A Case Study from the German Chemical Industry," *Isis* 73 (1982): 363–81.

19. Hunt to Vaughan, July 10, 1922, NRC, MED: Projects 1922, Pharmacology Studies; Proposed folder.

20. E. K. Marshall Jr., "Reid Hunt" *Biographical Memoirs* (Washington, D.C.: National Academy of Sciences) 26 (1951): 25–44.

21. Daniel P. Jones, "Chemical Warfare Research During World War I: A

Model of Cooperative Research," in *Chemistry and Modern Society: Historical Essays in Honor of Aaron J. Ihde,* ed. John Parascandola and James C. Whorton (Washington, D.C.: American Chemical Society, 1983), 165–85.

22. Robert E. Kohler, *From Medical Chemistry to Biochemistry: The Making of a Biomedical Discipline* (Cambridge: Cambridge University Press, 1982); Gerald L. Geison, "Divided We Stand: Physiologists and Clinicians in the American Context," and Russell C. Maulitz, "'Physician Versus Bacteriologist': The Ideology of Science in Clinical Medicine," both in *The Therapeutic Revolution: Essays in the Social History of American Medicine,* ed. Morris J. Vogel and Charles E. Rosenberg (Philadelphia: University of Pennsylvania Press, 1979).

23. Jonathan Liebenau, *Medical Science and Medical Industry: The Formation of the American Pharmaceutical Industry* (Baltimore: Johns Hopkins University Press, 1987); Swann, *Academic Scientists.* In the context of pharmaceutical manufacture in the nineteenth and early twentieth centuries, "ethical" referred to manufacturers whose products were marketed chiefly to physicians rather than directly to consumers.

24. Parascandola, *Development of American Pharmacology,* 116–25; David A. Hounshell, "The Evolution of Industrial Research in the United States," in *Engines of Innovation: U.S. Industrial Research at the End of an Era,* ed. Richard S. Rosenbloom and William J. Spencer (Boston: Harvard Business School Press, 1996), 13–85.

25. Parascandola, *Development of American Pharmacology,* 1.

26. Quoted in ibid., 44.

27. See, e.g., W. T. Councilman, W. W. Grant, and M. L. Harris, "Special Report on the Council on Pharmacy and Chemistry," *American Medical Association Bulletin* 10 (15 May 1915): 338; and Alfred Newton Richards, "The Opium Problem," "The Scope of the Committee," etc. (typescripts, n.d.), ANR, box 11, folder 41.

28. Hunt to Vaughan, July 10, 1922, NRC, MED: Projects 1922–Pharmacology Studies; Proposed folder.

29. Charles E. Rosenberg, "The Therapeutic Revolution: Medicine, Meaning, and Social Change in Nineteenth-century America," in *The Therapeutic Revolution: Essays in the Social History of American Medicine,* ed. Morris J. Vogel and Charles E. Rosenberg (Philadelphia: University of Pennsylvania Press, 1979), 3–25.

30. Virginia Berridge and Griffith Edwards, *Opium and the People: Opiate Use in Nineteenth-Century England* (New York: St. Martin's Press, 1981).

31. James G. Burrow, *A.M.A.: Voice of American Medicine* (Baltimore: Johns Hopkins Press, 1963).

32. On the Council on Pharmacy and Chemistry's activities, see Harry Marks,

The Progress of Experiment: Science and Therapeutic Reform in the United States, 1900–1990 (Cambridge: Cambridge University Press, 1997).

33. Charles E. Terry and Mildred Pellens, *The Opium Problem* (New York: Bureau of Social Hygiene, 1928).

34. Robert E. Kohler, *Partners in Science: Foundations and Natural Scientists 1900–1945* (Chicago: University of Chicago Press, 1991).

35. Adele E. Clarke, *Disciplining Reproduction: Modernity, American Life Sciences, and the Problems of Sex* (Berkeley: University of California Press, 1998), esp. 90–118.

36. David T. Courtwright, "Introduction: The Classic Era of Narcotic Control," in *Addicts Who Survived: An Oral History of Narcotic Use in America, 1923–1965*, ed. id., Herman Joseph, and Don Des Jarlais (Knoxville: University of Tennessee Press, 1989), 1–44; Jim Baumohl, "Maintaining Orthodoxy: The Depression-Era Struggle over Morphine Maintenance in California," in *Altering American Consciousness: Essays on the History of Alcohol and Drug Use in the United States, 1800–2000*, ed. Sarah W. Tracy and Caroline Jean Acker (Amherst: University of Massachusetts Press, forthcoming).

37. Minutes of meeting, Committee on Drug Addiction, May 3, 1929, NRC, Meetings—Minutes—1929 folder. On Small's career, see Erich Mosettig, "Lyndon Frederick Small," *Biographical Memoirs* (Washington, D.C.: National Academy of Sciences) 33 (1959): 397–413.

38. "Pharmacological Research, Committee on" (index cards), NRC, MED: Activities 1919–23, Administrative Summary folder.

39. Richards to Matthews, 20 Nov. 1933, ANR, box 11, folder 27.

40. White to Dunham 29 Jan. 1929, NRC, MED: Com on Drug Addiction 1928–1929; Beginning of Program folder.

41. White to Kellogg, June 20, 1929 (memo); White to E. V. Cowdry, Jan. 5, 1930, both in NRC MED: Com on Drug Addiction 1928–1930, Appointments: Members folder.

42. "Conference on Patents; Committee on Drug Addiction; February 10, 1937 Transcript of Discussion," NRC, LEGAL Matters: Patents 1936–1937, Policy re Com on Drug Addiction Patents, Conference on Disposition of Narcotics Patents: Joint Com on Patent Policy folder.

43. Marks, *Progress of Experiment*, 24, 34.

44. Report of the chairman, Committee on Drug Addiction, Sept. 1, 1931, NRC, Committee on Drug Addiction Meeting Minutes folder.

45. "Minutes of the Committee on Drug Addiction," Nov. 3, 1929, NRC, Meetings—Minutes—1929 folder. The book was *The Indispensable Use of Narcotics* (Chicago: American Medical Association, 1931).

46. Small, ca. June 1937, memorandum to committee (n.d.), NRC, LEGAL

Matters: Patents 1933–1938, Small LF: Morphine Derivatives & Processes for the Preparation folder.

47. "Conference on Patents . . . February 10, 1937, Transcript of Discussion," NRC, LEGAL Matters: Patents 1936–1937, Policy re Com on Drug Addiction Patents, Conference on Disposition of Narcotics patents: Joint Com on Patent Policy folder.

48. Small to Randolph Major, Mar. 25, 1935, and May 11, 1936, both in LFS, box 3, ME 1935–36 folder.

49. George Merck to Small, Apr. 24, 1934, LFS, box 3, ME 1934–35 folder; Mrs. Lyndon F. Small to William Charles White, Dec. 7, 1936, and White to Small, Dec. 24, 1936, both in LFS, box 4, WHITE 1935–39 folder.

50. Small to Rosin, June 21, 1933, LFS, box 3, ME 1932–34 folder. Small also shared unpublished reports on dihydrodesoxymorphine-D's physiological activity with officials at Mallinckrodt; Small to Russe, June 14, 1934, LFS, box 2, MA 1933–34 folder.

51. Eli Lilly to Small, Nov. 16, 1936, LFS, box 2, folder L.

52. Correspondence between Small and various individuals at Merck, LFS, box 3, ME 1934–35 folder.

53. Rosin to Small, June 28, 1933, LFS, box 3, ME 1932–34 folder.

54. Bayne-Jones to Charles Edmunds, Mar. 24, 1933, NRC, LEGAL Matters: Patents 1933–1938—Policy re Com on Drug Addiction Patents—General folder.

55. Charles Weiner, "Patents and Academic Research," in *Owning Scientific and Technical Information: Value and Ethical Issues,* ed. Vivian Weil and John W. Snapper (New Brunswick, N.J.: Rutgers University Press, 1989), 87–109.

56. Paul Nicholas Leech to Small, Feb. 20, 1935, box 2, folder L; Edmunds to Small, Feb. 4 and 23, 1935, box 2, EC-EI folder; Small to Joseph Rosin, Feb. 28, 1935, box 3, ME 1935–36 folder; Small to White, Feb. 28, 1935, and White to Small, Mar. 6, 1935, box 4, WHITE 1935–36 folder; Rosin to Small, Mar. 8, 1935, box 3, ME 1935–36 folder; all in LFS.

57. For a survey of Clifton K. Himmelsbach's career and his recollections of his work with addicted subjects in the 1930s, see his Oral History Interview at the History of Medicine Division, National Library of Medicine, Bethesda, Md.

58. William Charles White, "Preliminary Annual Report," Oct. 5, 1933, LFS, box 1, DO-DR folder.

59. Nathan B. Eddy and C. K. Himmelsbach, "Experiments on the Tolerance and Addiction Potentialities of Dihydrodesoxymorphine-D ('Desomorphin')," *Public Health Reports* Supplement 118 (1936): 2.

60. Robert Hatcher, "Report to the Committee on Drug Addiction of the National Research Council," Dec. 27, 1930, NRC, Committee on Drug Addiction Meetings: Minutes folder.

61. William Charles White, "Report of the Chairman, December 27, 1930," NRC, Meetings—Minutes—1930–32 folder.

62. Author's interview with Clifton K. Himmelsbach, Bethesda, Md., Oct. 28, 1993.

63. William Charles White, "Report of the Committee on Drug Addiction, October, 1934," LFS, box 3, NA 1934–35 folder; author's interview with Clifton K. Himmelsbach, Bethesda, Md., Oct. 28, 1993.

64. Himmelsbach to Treadway, Apr. 1 and 16, 1935, both in LFS, box 4, TREASURY DEPT. 1935–39 folder.

65. Edmunds to Small, Apr. 16, 1935, and Small to Edmunds, May 28, 1935, LFS, box 1, Correspondence A-E folder.

66. Eddy and Himmelsbach, "Experiments on . . . Dihydrodesoxymorphine-D."

67. M. R. King, C. K. Himmelsbach, and B. S. Sanders, "Dilaudid (Dihydromorphinone): A Review of the Literature and a Study of Its Addictive Properties," *Public Health Reports* Supplement 113 (1935).

68. Memorandum from L. F. Small, Feb. 3, 1936, NRC, LEGAL Matters: Patents 1933–1938, Small LF: Morphine Derivatives & Processes for their Preparation folder.

69. Himmelsbach to White, Jan. 17, 1938, NRC, LEGAL Matters: Patents 1933–1938, Small LF: Morphine Derivatives and Processes for their Preparation folder.

70. See *The Right Tools for the Job: At Work in Twentieth-Century Life Sciences,* ed. Adele Clarke and Joan Fujimura (Princeton: Princeton University Press, 1992). The essays in this collection examine what factors make a given organism "right" for particular kinds of laboratory investigation.

71. Abraham Lilienfeld, "*Ceteris Paribus:* The Evolution of the Clinical Trial," *Bulletin of the History of Medicine* 56 (1982): 1–18; Marks, *Progress of Experiment.*

72. Marks, *Progress of Experiment,* 49 ff.

73. Arthur E. Jacobson and Kenner C. Rice, "The Continuing Interrelationship of CPDD and NIDDK," in *Problems of Drug Dependence: 1991,* ed. Louis S. Harris, National Institute on Drug Abuse Research Monograph (Washington, D.C.: GPO, 1992); Mosettig, "Lyndon Frederick Small." Small and Eddy worked on anti-malarials until the end of World War II; they then resumed their work on opiates for the duration of their careers.

74. Nathan B. Eddy, *The National Research Council Involvement in the Opiate Problem, 1928–1971* (Washington, D.C.: National Academy of Sciences, 1973), 43–44.

75. Louis Goodman and Alfred Gilman, *The Pharmacological Basis of Therapeutics* (New York: Macmillan, 1941).

76. "Report on Clinical Studies Particularly in Respect to Metopon, June 19 to 26, 1939, inclusive" (n.d.), MED: Com on Drug Addiction 1939 folder; NRC, Projects: Development of Nonaddictive Analgesics: Clinical Studies Conferences.

77. Clifton K. Himmelsbach, "Studies of Certain Addiction Characteristics of (a) Dihydromorphine ('Paramorphan'), (b) Dihydrodesoxymorphine-D ('Desomorphine'), (c) Dihydrodesoxycodeine-D ('Desocodeine'), and (d) Methyldihydromorphinone ('Metopon')," *Journal of Pharmacology and Experimental Therapeutics* 67, no. 2 (Oct. 1939): 239–49.

78. Lyndon E. Lee, "Studies of Morphine, Codeine and Their Derivatives. XVI. Clinical Studies of Morphine, Methyldihydromorphinone (Metopon) and Dihydrodesoxymorphine-D (Desmorphine [*sic*])," *Journal of Pharmacology and Experimental Therapeutics* 75 (1942): 161–73.

79. Alfred Newton Richards, "The Opium Problem," "The Scope of the Committee," etc. (typescripts, n.d.), ANR, box 11, folder 41.

80. All quotations from ibid. Emphasis in original.

81. Solomon H. Snyder, *Brainstorming: The Science and Politics of Opiate Research* (Cambridge, Mass.: Harvard University Press, 1989).

82. Kohler, *Partners in Science.*

83. Glenn E. Bugos, "Managing Cooperative Research and Borderland Science in the National Research Council, 1922–1942," *Historical Studies in the Physical Sciences* 20, no. 1 (1989): 1–32.

84. Clarke, *Disciplining Reproduction.*

85. Washington *Post,* Nov. 9, 1936 (copy in NRC, PUBS 1931–1939 folder); correspondence among Eddy, E. M. K. Geiling, White, and Anslinger, Dec. 4 to 22, 1929, NRC, MED: Com on Drug Addiction 1929–1939, Grantees folder.

86. Himmelsbach, Oral History, National Library of Medicine.

Chapter 4. Constructing the Addict Career

1. PCCSOA, vol. 1, case 26–32. Patient names have been changed and minor errors of transcription have been corrected.

2. Elizabeth Lunbeck, *The Psychiatric Persuasion: Knowledge, Gender, and Power in Modern America* (Princeton: Princeton University Press, 1994), 132.

3. Arthur B. Light et al., *Opium Addiction* (Chicago: American Medical Association, 1929–30).

4. Caroline J. Acker, "From All-Purpose Anodyne to Marker of Deviance: Physicians' Attitudes Toward Opiates in the U.S., 1890–1940," in *Drugs and*

Narcotics in History, ed. Roy Porter and Mikulas Teich (Cambridge: Cambridge University Press, 1995), 126–27.

5. Ruth Rosen, *The Lost Sisterhood: Prostitution in America, 1900–1918* (Baltimore: Johns Hopkins University Press, 1982), 102.

6. Paul G. Cressey, *The Taxi-Dance Hall: A Sociological Study in Commercialized Recreation and City Life* (Chicago: University of Chicago Press, 1932; reprint, New York: Greenwood Press, 1968), 54.

7. Light et al., *Opium Addiction.*

8. PCCSOA, vol. 3, case 26–62.

9. PCCSOA, vol. 5, case 26–4.

10. Howard S. Becker, *Outsiders: Studies in the Sociology of Deviance* (New York: Free Press, 1963).

11. Charles E. Faupel, *Shooting Dope: Career Patterns of Hard-Core Heroin Users* (Gainesville: University of Florida Press, 1991).

12. Joseph reported using 9–12 grains per day in the period just before his current admission at Philadelphia General Hospital, and he reported spending between $30 and $35 an ounce for heroin, a typical price in that period. Since an ounce contains 480 grains, an ounce was lasting him between about 40 and 53 days. At $30 an ounce, this meant at least $22.50 monthly. His wages amounted to about $120 a month (assuming an average of slightly more than four weeks).

13. PCCSOA, vol. 4, case 26–79.

14. Cressey, *Taxi-Dance Hall.*

15. PCCSOA, vol. 11A, case 26–33.

16. PCCSOA, vol. 21, case 26–82.

17. Marsha Rosenbaum, *Women on Heroin* (New Brunswick, N.J.: Rutgers University Press, 1981).

18. Jack D. Pressman, *Last Resort: Psychosurgery and the Limits of Medicine* (Cambridge: Cambridge University Press, 1998).

19. Lunbeck, *Psychiatric Persuasion.*

20. Ibid., 306, 65–71; Frederic M. Thrasher, *The Gang: A Study of 1,313 Gangs in Chicago* (Chicago: University of Chicago Press, 1927), 507.

21. Charles Terry to Alfred Newton Richards, Dec. 31, 1926, and Nov. 2, 1927, ANR, box 11, folder 24.

22. Minutes, CDA meeting of Dec. 17, 1927, ANR, box 11, folder 29.

23. Pressman, *Last Resort.*

24. Remarks by Joseph Doane, "Chemical and Clinical Studies of Opium Addiction. Program given before the Philadelphia County Medical Society," Dec. 19, 1928 (typescript), ANR, box 22, folder 29.

25. Minutes, CDA meeting of Dec. 17, 1927, ANR, box 11, folder 29.

26. Correspondence in ANR, box 11, folder 26.

27. Minutes, CDA meeting of Oct. 7, 1928, box 11, folder 129.

Chapter 5. The Junkie as Psychopath

1. For discussions of Kolb, see David Musto, *The American Disease: Origins of Narcotic Control* (New Haven: Yale University Press, 1973); David T. Courtwright, *Dark Paradise: Opiate Addiction in America Before 1940* (Cambridge, Mass.: Harvard University Press, 1982); H. Wayne Morgan, *Drugs in America: A Social History, 1800–1980* (Syracuse, N.Y.: Syracuse University Press, 1981); and Arnold Jaffe, *Addiction Reform in the Progressive Age: Scientific and Social Responses to Drug Dependence in the U.S., 1870–1930* (New York: Arno Press, 1981).

2. In using the term "insane," I follow the usage of the nineteenth-century actors I discuss.

3. Paula S. Fass, *The Damned and the Beautiful: American Youth in the 1920s* (Oxford: Oxford University Press, 1977).

4. John Burnham, *Bad Habits: Drinking, Smoking, Taking Drugs, Gambling, Sexual Misbehavior, and Swearing in American History* (New York: New York University Press, 1993).

5. Dashiell Hammett, *The Thin Man* (1932; New York: Knopf, 1934).

6. W. S. Van Dyke, director, *The Thin Man* (Metro-Goldwyn-Mayer, 1934).

7. David T. Courtwright, "Charles Terry, *The Opium Problem,* and American Narcotic Policy," *Journal of Drug Issues* 16 (1986): 421–34, p. 422. Charles Terry had graduated from the University of Maryland School of Medicine in 1903.

8. Biographical notes written in Kolb's hand, December 1912, KP, box 1, Biographical Data folder.

9. Ralph Chester Williams, *The United States Public Health Service, 1798–1950* (Washington, D.C.: USPHS, 1951); Victoria A. Harden, *Inventing the NIH: Federal Biomedical Research Policy, 1887–1937* (Baltimore: Johns Hopkins University Press, 1986).

10. For a discussion of the physical and mental health screening at Ellis Island, see Alan M. Kraut, *Silent Travelers: Germs, Genes, and the "Immigrant Menace"* (Baltimore: Johns Hopkins University Press, 1994), 51–75.

11. On the transformation of American hospitals from such institutions to ones where medical concerns dominated, see Charles E. Rosenberg, *The Care of Strangers: The Rise of America's Hospital System* (New York: Basic Books, 1987); David Rosner, *A Once Charitable Enterprise: Hospitals and Health Care in Brooklyn and New York, 1885–1915* (New York: Cambridge University Press, 1982; reprint, Princeton: Princeton University Press, 1987); and Morris J. Vogel, *The*

Invention of the Modern Hospital, Boston, 1870–1930 (Chicago: University of Chicago Press, 1980).

12. Nancy Tomes, *A Generous Confidence: Thomas Story Kirkbride and the Art of Asylum-Keeping, 1840–1883* (Cambridge: Cambridge University Press, 1984); Gerald Grob, *Mental Illness and American Society, 1875–1940* (Princeton: Princeton University Press, 1983).

13. John B. Chapin, "Presidential Address," *American Journal of Insanity* 46 (1889): 12–13.

14. Erwin H. Ackerknecht, *Medicine at the Paris Hospital, 1794–1848* (Baltimore: Johns Hopkins Press, 1967); Richard Shryock, *The Development of Modern Medicine: An Interpretation of the Social and Scientific Factors Involved* (Madison: University of Wisconsin Press, 1974, 1936); Michel Foucault, *The Birth of the Clinic: An Archaeology of Medical Perception,* trans. A. M. Sheridan Smith (New York: Vintage Books, 1975, 1994).

15. See, e.g., Edward Cowles, "The Mechanism of Insanity. Part II. The Normal Mechanism in Use," *American Journal of Insanity* 47 (1891): 471–95.

16. "Proceedings of the Association," *American Journal of Insanity* 47 (1890): 166–239.

17. "Proceedings of the Association," *American Journal of Insanity* 48 (1891): 105.

18. Grob, *Mental Illness,* 61–62.

19. Walter Channing, "Some Remarks on the Address . . . by S. Weir Mitchell," *American Journal of Insanity* 51 (1894): 171–81.

20. Spitska, "Letter to S. Weir Mitchell," *American Journal of Insanity* 51 (1894): 102–3.

21. August Hoch, "Kraepelin on Psychological Experimentation in Psychiatry," *American Journal of Insanity* 52 (1896): 387–96.

22. Frederick Peterson, "Some Problems of the Alienist," *American Journal of Insanity* 56 (1899): 1–19.

23. Jack D. Pressman, *Last Resort: Psychosurgery and the Limits of Medicine* (Cambridge: Cambridge University Press, 1998), 18–47; id., "Inventing Psychiatry" (unpublished MS, 1993); Elizabeth Lunbeck, *The Psychiatric Persuasion: Knowledge, Gender, and Power in Modern America* (Princeton: Princeton University Press, 1994).

24. Lunbeck, *Psychiatric Persuasion,* 65–71.

25. Kraut, *Silent Travelers,* 74.

26. This screening effort is described in Pearce Bailey, Frankwood E. Williams, and Paul O. Komora, "In the United States," in Surgeon General of the Army, *The Medical Department of the United States Army in the World War,* vol. 10: *Neuropsychiatry* (Washington, D.C.: GPO, 1929).

27. *Neuropsychiatry* (cited n. 26 above), 1–2.

28. Ibid., 8.

29. Ibid., 30, 33.

30. Ibid., 9.

31. Ibid., 73.

32. Ibid., 64–65.

33. Ibid., 67.

34. Ibid., 69.

35. Ibid., 257.

36. Ibid., 256–57.

37. Autobiography, enclosed with Kolb to surgeon general, Sept. 9, 1927, KP, box 1, Biographical Data folder.

38. Lawrence Kolb and A. G. DuMez, "Prevalence and Trend of Drug Addiction in the U.S. and Factors Influencing It," *Public Health Reports* 39 (1924): 1179–1204.

39. On estimates of addiction prevalence in this period, see Courtwright, *Dark Paradise,* 9–34.

40. Charles E. Terry and Mildred Pellens, *The Opium Problem* (New York: Bureau of Social Hygiene, 1928).

41. Lawrence Kolb, "Types and Characteristics of Drug Addicts," *Mental Hygiene* 9 (Apr. 1925): 300–313; "Drug Addiction in Its Relation to Crime," *Mental Hygiene* 9 (Jan. 1925): 74–89; "Pleasure and Deterioration from Narcotic Addiction," *Mental Hygiene* 9 (Oct. 1925): 699–724; "Relation of Intelligence to Etiology of Drug Addiction," *American Journal of Psychiatry* 5 (July 1925): 163–67; and "Addiction: Medical Cases," *Archives of Neurology and Psychiatry* 20 (July 1928): 171–83.

42. Allan M. Brandt, *No Magic Bullet: A Social History of Venereal Disease in the United States Since 1880* (1985; rev. ed., New York: Oxford University Press, 1987).

43. Kolb, "Types and Characteristics," 303.

44. Kolb, "Drug Addiction and Its Relation to Crime," 88–89.

45. Morgan, *Drugs in America,* 129; Ellen Fitzpatrick, *Endless Crusade: Women Social Scientists and Progressive Reform* (New York: Oxford University Press, 1990), 123; Lunbeck, *Psychiatric Persuasion,* 43, 62–63, 68–69, and 308.

46. McCoy to Stimson, Jan. 5, 1925 (memorandum written by Kolb), KP, box 5, Correspondence Sti–Sw folder. Courtwright, *Dark Paradise,* describes the close links between Kolb's medical arguments and the policy objectives of the Treasury Department.

47. Kolb, "Re desirability of use of propaganda to acquaint the masses of con-

sequences of using dangerous drugs" (typescript), July 28, 1925, KP, box 4, McCoy, G. W., folder.

48. McCoy to Stimson, cited n. 47 above.

49. See KP, box 7, for Kolb's discussion of these cases.

50. "Case of K. J. L. . . . Examined September 18, 1926," KP, box 7, Narcotics' Case Histories—August 1926 to 1928 folder.

51. "Case of R. T. E. . . . Examined March 26, 1926," KP, box 7, Narcotics' Case Histories January to June 1926 folder.

52. "Case of N. S. F. . . . Examined on January 26, 1926," KP, box 7, Narcotics' Case Histories January to June 1926 folder.

53. "Case of W. C. B. . . . Examined on September 7, 1926," KP, box 7, Narcotics' Case Histories—August 1926 to 1928 folder.

54. Kolb to Levi Nutt, June 25, 1928, KP, box 7, Narcotics Case Histories—August 1926 to 1928 folder. Only a brief report by Kolb is present; there is no history of the individual comparable to those provided for most other cases in the series.

55. "Let's Stop This Narcotics Hysteria," *Saturday Evening Post,* July 28, 1956, 19, 50, 54–55.

56. Kolb to Dr. C. A. D. Fairfield, July 27, 1927, KP, box 3, Correspondence E–Fel folder.

57. "H. J. Resolution 78: inves. narcotics and methods of treating addicts" (memorandum), Dec. 13, 1927, KP, box 4, McCoy, G. W., folder. By "treatment" in this context, Kolb simply meant detoxification.

Chapter 6. Healing Vision and Bureaucratic Reality

1. Jim Baumohl, "Maintaining Orthodoxy in California," in *Altering American Consciousness,* ed. Sarah W. Tracy and Caroline Jean Acker (Amherst: University of Massachusetts Press, forthcoming); Jim Baumohl and Sarah W. Tracy, "Building Systems to Manage Inebriates: The Divergent Paths of California and Massachusetts, 1891–1920," *Contemporary Drug Problems* (1994): 557–97.

2. David Musto, *The American Disease: Origins of Narcotic Control* (New Haven: Yale University Press, 1973), 204.

3. On the Public Health Service Narcotic Hospital at Lexington, Kentucky, see National Archives and Records Administration, National Archives, Washington, D.C., Records of the Public Health Service, 1912–1968 (Record Group 90) and records at NARA's Southeast Region facility in Atlanta, as well as some correspondence in the Lawrence Kolb papers at the National Library of Medicine and the Bureau of Social Hygiene records at the Rockefeller Archive Center. These records support an analysis of how the Lexington Narcotic Hospital functioned

as an institution in the eyes of its managers but do not contain material on the experience of being an inmate. For reminiscences of time spent at Lexington, see *Addicts Who Survived: An Oral History of Narcotic Use in America,* ed. David T. Courtwright, Herman Joseph, and Don Des Jarlais (Knoxville: University of Tennessee Press, 1989), 296–318.

4. Guenter B. Risse, *Mending Bodies, Saving Souls: A History of Hospitals* (New York: Oxford University Press, 1999).

5. Baumohl and Tracy, "Building Systems," 561.

6. Alan M. Kraut, *Silent Travelers: Germs, Genes, and the "Immigrant Menace"* (Baltimore: Johns Hopkins University Press, 1994), 160.

7. William L. White, *Slaying the Dragon: The History of Addiction Treatment and Recovery in America* (Bloomington, Ill.: Chestnut Hill Systems, 1998); Sarah W. Tracy, "The Foxborough Experiment: Medicalizing Inebriety at the Massachusetts Hospital for Dipsomaniacs and Inebriates, 1833 to 1919" (Ph.D. diss., University of Pennsylvania, 1992).

8. David Rothman, *Conscience and Convenience: The Asylum and Its Alternatives in Progressive America* (Boston: Little, Brown, 1980).

9. Rothman, *Conscience and Convenience,* 117–58.

10. Musto, *American Disease,* 204–6; H. Wayne Morgan, *Drugs in America: A Social History, 1800–1980* (Syracuse, N.Y.: Syracuse University Press, 1981), 135–36.

11. Cited in Jack D. Pressman, *Last Resort: Psychosurgery and the Limits of Medicine* (Cambridge: Cambridge University Press, 1998), 222; see also 156, 162, 218–23.

12. Ralph Chester Williams, *The United States Public Health Service, 1798–1950* (Washington, D.C.: USPHS, 1951), 168.

13. Ibid., 146, 334–35, 377.

14. John D. Farnham, file memo, Dec. 18, 1931, RAC, Bureau of Social Hygiene collection, ser. 3, subser. 1, box 3, folder 131. The Rotary Clubs are mentioned in Treadway to Dunham, Feb. 23, 1932, ibid., box 4, folder 132.

15. Treadway to Fuller, Apr. 24, 1935, NA, box 107, 1960– (1935–).

16. Cumming's remarks are reproduced in a press release dated May 25, 1935, in NA, box 104, Dr. Cumming's Address at Dedication.

17. Dorothy Ross discusses the emergence of the idea of social control as a necessary means of sustaining cohesion in a complex industrial society in *The Origins of American Social Science* (Cambridge: Cambridge University Press, 1991), 229–40.

18. Secretary of the Treasury Henry Morgenthau Jr. to Speaker of the House, Dec. 13, 1935 (annual fiscal report), NA, box 103, 0115 Accts., Except, 1934–35, folder.

19. Pressman, *Last Resort,* 221.

20. Joel Braslow, *Mental Ills and Bodily Cures: Psychiatric Treatment in the First Half of the Twentieth Century* (Berkeley: University of California Press, 1997), 27–30, 36–37, 43–51, 104–7.

21. Pressman, *Last Resort,* passim.

22. NA boxes 106–8 contain reams of such requisitions.

23. Lawrence Kolb, "Report on the First United States Narcotic Farm, Lexington, Kentucky for the National Emergency Council," KP, box 5, San–Sh folder.

24. Lawrence Kolb and W. F. Ossenfort, "The Treatment of Drug Addicts at the Lexington Hospital," *Southern Medical Journal* 8, no. 31 (Aug. 1938): 914–22.

25. Ibid., 915.

26. Lawrence M. Friedman, *Crime and Punishment in American History* (New York: Basic Books, 1993), 425–27.

27. Ruth Rosen, *The Lost Sisterhood: Prostitution in America, 1900–1918* (Baltimore: Johns Hopkins University Press, 1982), 19.

28. "Monthly Report of Medical Activities of the First U.S. Narcotic Farm for the Month Ending September 30, 1935," NA, box 106, folder 1851.

29. Kolb and Ossenfort, "Treatment of Drug Addicts," 915, 918.

30. Pressman, *Last Resort,* 163.

31. J. R. Dawson, senior dairy husbandman, Bureau of Dairy Industry, Department of Agriculture, memorandum, Mar. 18, 1935, NA, box 108, 1975-152-75 folder.

32. Kolb to Treadway, June 5, 1935, NA, box 103, 0115, Accts., Except, 1934–35, folder.

33. Cumming to secretary of the Treasury, Feb. 2, 1925, NA, box 106, folder 0957.

34. Kolb to Treadway, Aug. 30, 1935, NA, box 105, folder 0605–0615.

35. "Monthly Report of Medical Activities of the First U.S. Narcotic Farm for the Month Ending September 30, 1935," NA, box 106, folder 1851.

36. Michael J. Pescor, "A Statistical Analysis of the Clinical Records of Hospitalized Drug Addicts," *Public Health Reports* Supplement 143 (Washington, D.C.: GPO, 1938), 17.

37. "Statistical Report for Period Ending September 30, 1935," NA, box 106, folder 1851.

38. Pescor, "Statistical Analysis," 11.

39. The occupations are listed in "Statistical Report for Period Ending September 30, 1935," NA, box 106, folder 1851.

40. Treadway to Anslinger, Oct. 20, 1932, NA, box 106, folder 1616.

41. Bates to Treadway, July 23, 1935, NA, box 106, folder 1616.

42. Anslinger to Treadway, Aug. 24, 1933; Anslinger to Cumming, Oct. 26, 1933; both in NA, box 106, folder 1616.

43. Kolb to surgeon general, Aug. 26, 1935; Kolb to Treadway, Aug. 28, 1935; "Testimony of Employees Regarding the Escape of the Four Voluntary Patients"; all in NA, box 106, folder 1616.

44. Kolb to Treadway, Oct. 24, 1935, NA, box 106, folder 1616.

45. Reed to secretary of the Treasury, Nov. 26, 1935, NA, box 106, folder 1616.

46. M. J. Pescor, "The Kolb Classification of Drug Addicts," *Public Health Reports* Supplement 155 (Washington, D.C.: GPO, 1939). The same 1,036 cases provided the basis for Pescor's "Statistical Analysis" article, cited above.

47. Arnold Jaffe, *Addiction Reform in the Progressive Age: Scientific and Social Responses to Drug Dependence in the U.S., 1870–1930* (New York: Arno Press, 1981), 239.

48. Baumohl, "Maintaining Orthodoxy."

49. Elizabeth Lunbeck, *The Psychiatric Persuasion: Knowledge, Gender, and Power in Modern America* (Princeton: Princeton University Press, 1994), 67.

50. Pressman, *Last Resort.*

51. Williams, *Public Health Service,* 52, 336, 354.

52. Ibid., 58.

53. Karst J. Besteman, "Federal Leadership in Building the National Drug Treatment System," in *Treating Drug Problems,* ed. Dean R. Gerstein and Henrick J. Harwood (Washington, D.C.: National Academy Press, 1992), 2: 63–88.

54. John A. O'Donnell, "The Relapse Rate in Narcotic Addiction: A Critique of Follow-Up Studies," in *Narcotics,* ed. D. M. Wilner and G. G. Kassebaum (New York: McGraw-Hill, 1965).

55. Rothman, *Conscience and Convenience.*

Chapter 7. The Addict in the Social Body

1. Talcott Parsons, *The Social System* (New York: Free Press, 1951).

2. Howard S. Becker, interview by the author, Mar. 4, 2000, Santa Barbara, California. Unless otherwise noted, information about Becker is from this interview.

3. Alfred R. Lindesmith, *Opiate Addiction* (Bloomington, Ind.: Principia Press, 1947). This book was a revision of Lindesmith's University of Chicago dissertation, which was published earlier in a private edition as *The Nature of Opiate Addiction* (Chicago: University of Chicago Libraries, 1937).

4. Howard S. Becker, *Outsiders: Studies in the Sociology of Deviance* (New York: Free Press, 1963).

5. Ibid., 3, 6, 21, 34, 169.

6. Ibid., 3.

7. Ibid., 21, 34.

8. John C. McWilliams, *The Protectors: Harry J. Anslinger and the Federal Bureau of Narcotics, 1930–1962* (Newark: University of Delaware Press, 1990), 111, 117; K. A. Cuordileone, "'Politics in an Age of Anxiety': Cold War Political Culture and the Crisis in American Masculinity, 1949–1960," *Journal of American History* 87, no. 2 (Sept. 2000): 515–45.

9. Allan M. Brandt, *No Magic Bullet: A Social History of Venereal Disease in the United States since 1880* (1985; rev. ed., New York: Oxford University Press, 1987); Elizabeth Fee, "Sin Versus Science: Venereal Disease in Twentieth-Century Baltimore," in *AIDS: The Burdens of History,* ed. id. and Daniel M. Fox (Berkeley: University of California Press, 1988), 121–46.

10. Walter C. Reckless, *Vice in Chicago* (Chicago: University of Chicago Press, 1933; reprint, Montclair, N.J.: Patterson Smith, 1969).

11. Vice Commission of Chicago, *The Social Evil in Chicago: A Study of Existing Conditions with Recommendations* (Chicago: Gunthorp-Warren, 1911; reprint, New York: Arno Press/New York Times, 1970).

12. Reckless, *Vice in Chicago,* ix.

13. Ibid., 87; Frederic Thrasher, *The Gang: A Study of 1,313 Gangs in Chicago* (Chicago: University of Chicago Press, 1927), 452–86.

14. Reckless, *Vice in Chicago,* 8, 14.

15. Ruth Rosen, *The Lost Sisterhood: Prostitution in America, 1900–1918* (Baltimore: Johns Hopkins University Press, 1982).

16. Reckless, *Vice in Chicago,* 164–97.

17. Robert C. Bannister, *Sociology and Scientism: The American Quest for Objectivity, 1880–1940* (Chapel Hill: University of North Carolina Press, 1987).

18. Ellen Fitzpatrick, *Endless Crusade: Women Social Scientists and Progressive Reform* (New York: Oxford University Press, 1990) 41.

19. Faris and Dunham, *Mental Disorders in Urban Areas: An Ecological Study of Schizophrenia and Other Psychoses* (Chicago: University of Chicago Press, 1939); R. E. Park and E. W. Burgess, *The City* (Chicago: University of Chicago Press, 1925).

20. Bannister, *Sociology and Scientism;* Dorothy Ross, *The Origins of American Social Science* (Cambridge: Cambridge University Press, 1991).

21. The phrase is Ross's: *Origins of American Social Science,* 308.

22. Ibid., 226, 307–8, 401.

23. Paul G. Cressey, *The Taxi-Dance Hall: A Sociological Study in Commercialized Recreation and City Life* (Chicago: University of Chicago Press, 1932; reprint, New York: Greenwood Press, 1968).

24. See, e.g., Paul Colomy and J. David Brown, "Elaboration, Revision, Polemic, and Progress in the Second Chicago School," in *A Second Chicago School? The Development of a Postwar Sociology,* ed. Gary Alan Fine (Chicago: University of Chicago Press, 1995), 27; Ross, *Origins of American Social Science,* 311.

25. Cressey, *Taxi-Dance Hall,* 3.

26. Ibid., 126.

27. Ibid., ch. 7, "The Filipino and the Dance Hall."

28. Ibid., 130.

29. Ibid., 141.

30. Ibid., 143.

31. Ibid., 7.

32. Pearce Bailey, "The Heroin Habit," *New Republic* 6 (Apr. 22, 1916): 314–16, reprinted in *Yesterday's Addicts: American Society and Drug Abuse, 1865–1920,* ed. H. Wayne Morgan (Norman: University of Oklahoma Press, 1974), 171–76.

33. Harvey W. Feldman and Michael R. Aldrich, "The Role of Ethnography in Substance Abuse Research and Public Policy: Historical Precedent and Future Prospects," in *The Collection and Interpretation of Data from Hidden Populations,* ed. Elizabeth Y. Lambert, NIDA Research Monograph 98 (Rockville, Md.: National Institute on Drug Abuse, 1990), 12–30.

34. Bingham Dai, *Opium Addiction in Chicago* (Shanghai: Commercial Press, 1937), v.

35. Edwin Sutherland to Lawrence Dunham, Nov. 15, 1933, RAC, Bureau of Social Hygiene collection, ser. 3, subser. 1, box 4, folder 134.

36. Bingham Dai, "A Project for the Sociological Study of Opium Addiction" (typescript, n.d.), RAC, Bureau of Social Hygiene collection, ser. 3, subser. 1, box 4, folder 134.

37. Sutherland to Dunham, Nov. 15, 1933, RAC, Bureau of Social Hygiene collection, ser. 3, subser. 1, box 4, folder 134.

38. Dunham to Sutherland, Nov. 17, 1933, RAC, Bureau of Social Hygiene collection, ser. 3, subser. 1, box 4, folder 134.

39. Dai, "Project for the Sociological Study," 1.

40. Dai, *Opium Addiction in Chicago,* vi.

41. Ibid., 73–94.

42. Ibid., 95.

43. Ross, *Origins of American Social Science,* 347–57.

44. Dai, *Opium Addiction in Chicago,* 96.

45. Ibid., 98.

46. Ibid., 14.

47. Ibid., 110.

48. Ibid., 10.

49. Ibid., 12.

50. For a description of this facility, see H. Wayne Morgan, *Drugs in America: A Social History, 1800–1980* (Syracuse, N.Y.: Syracuse University Press, 1981), 74–83, and William L. White, *Slaying the Dragon: The History of Addiction Treatment and Recovery in America* (Bloomington, Ill.: Chestnut Hill Systems, 1998).

51. David Patrick Keys and John F. Galliher, *Confronting the Drug Control Establishment: Alfred Lindesmith as a Public Intellectual* (Albany: State University of New York Press, 2000), 70.

52. Lindesmith, *Opiate Addiction.*

53. A. R. Lindesmith, "A Sociological Theory of Drug Addiction," *American Journal of Sociology* 43 (1938): 593–613.

54. Kolb to Victor Vogel, Jan. 20, 1938, KP, box 5, U–Vog folder.

55. Lawrence Kolb and A. G. DuMez, "Experimental Addiction of Animals to Opiates," *Public Health Reports,* Mar. 27, 1931, 698–726.

56. Alfred R. Lindesmith, *The Addict and the Law* (Bloomington: Indiana University Press, 1968).

57. Alfred R. Lindesmith and Anselm Strauss, *Social Psychology* (New York: Dryden Press, 1949).

58. Keys and Galliher, *Confronting the Drug Control Establishment,* 101–5.

59. Jim Baumohl, "Maintaining Orthodoxy in California," in *Altering American Consciousness,* ed. Sarah W. Tracy and Caroline Jean Acker (Amherst: University of Massachusetts Press, forthcoming); McWilliams, *Protectors;* Keys and Galliher, *Confronting the Drug Control Establishment.*

60. Robert K. Merton, "Social Structure and Anomie," in *Social Theory and Social Structure* (New York: Free Press, 1957).

61. William Charles White, "Committee on Drug Addiction" (n.d.), RAC, Bureau of Social Hygiene collection, ser. 1, box 5, folder 142.

62. Author's interview with Clifton K. Himmelsbach, Bethesda, Md., Oct. 28, 1993.

63. Walter B. Cannon, *The Wisdom of the Body* (1932; rev. ed., New York: Norton, 1939), 24.

64. Ibid., 305–6.

65. Ibid., 311–12.

66. Ibid., 317. Emphasis in original.

67. Becker, *Outsiders,* 14.

68. Feldman and Aldrich, "Role of Ethnography," 19.

69. Erving Goffman, *Stigma: Notes on the Management of Spoiled Identity* (New York: Touchstone, 1963).

70. Dai, *Opium Addiction in Chicago,* 97.

71. Lindesmith, *Opiate Addiction,* 11.

72. American Medical Association and the American Bar Association Joint Committee on Narcotic Drugs, *Drug Addiction: Crime or Disease* (Bloomington: Indiana University Press, 1961); David Musto, *The American Disease: Origins of Narcotic Control* (New Haven: Yale University Press, 1973), 232–33.

73. Feldman and Aldrich, "Role of Ethnography."

74. Edward Preble and John J. Casey, "Taking Care of Business—The Heroin User's Life on the Street," *International Journal of the Addictions* 4, no. 1 (1969): 1–24.

75. Philippe Bourgois, *In Search of Respect: Selling Crack in El Barrio* (Cambridge: Cambridge University Press, 1995).

76. Feldman and Aldrich, "Role of Ethnography," 22.

77. Patricia Cleckner, "Freaks and Cognoscenti: PCP Use in Miami," in *Angel Dust: An Ethnographic Study of PCP Users,* ed. Harvey W. Feldman, Michael H. Agar, and George M. Beschner (Lexington, Mass.: Lexington Books, 1980).

78. Patrick Biernacki, *Pathways from Heroin Addiction: Recovery Without Treatment* (Philadelphia: Temple University Press, 1986).

79. Marsha Rosenbaum, *Women on Heroin* (New Brunswick, N.J.: Rutgers University Press, 1981), 5.

80. Charles E. Faupel, *Shooting Dope: Career Patterns of Hard-Core Heroin Users* (Gainesville: University of Florida Press, 1991).

81. Talcott Parsons, "Definitions of Health and Illness in the Light of American Values and Social Structure," in *Concepts of Health and Disease: Interdisciplinary Perspectives,* ed. Arthur L. Caplan, H. Tristram Engelhardt, and J. J. McCarney (Reading, Mass.: Addison Wesley, 1981).

82. Caroline Jean Acker, "Stigma or Legitimation? A Historical Examination of the Social Potentials of Addiction Disease Models," *Journal of Psychoactive Drugs* 25, no. 3 (1993): 193–205.

83. Parsons, *Social System,* 250, 268, 341.

84. John A. O'Donnell, "The Relapse Rate in Narcotic Addiction: A Critique of Follow-Up Studies," in *Narcotics,* ed. D. M. Wilner and G. G. Kassebaum (New York: McGraw-Hill, 1965).

Conclusion

1. Caroline J. Acker, "From All-Purpose Anodyne to Marker of Deviance: Physicians' Attitudes Towards Opiates in the U.S., 1890–1940," in *Drugs and Narcotics in History,* ed. Roy Porter and Mikulas Teich (Cambridge: Cambridge University Press, 1995), 114–32.

2. David T. Courtwright, "Introduction: The Classic Era of Narcotic Control," in *Addicts Who Survived: An Oral History of Narcotic Use in America, 1923–1965*, ed. id., Herman Joseph, and Don Des Jarlais (Knoxville: University of Tennessee Press, 1989), 1–44.

3. David T. Courtwright, "The Prepared Mind: Marie Nyswander, Methadone Maintenance, and the Metabolic Theory of Addiction," *Addiction* 92, no. 3 (1997): 257–65.

4. Author's interview with David E. Smith, San Francisco, July 21, 2000.

5. Caroline Jean Acker, "Stigma or Legitimation? A Historical Examination of the Social Potentials of Addiction Disease Models," *Journal of Psychoactive Drugs* 25 (1993): 193–205.

6. Institute of Medicine, *Treating Drug Problems*, vol. 1, ed. Dean R. Gerstein and Henrick J. Harwood (Washington, D.C.: National Academy Press, 1990); Peter Goldberg, "The Federal Government's Response to Illicit Drugs, 1969–1978," in Drug Abuse Council, *The Facts about "Drug Abuse"* (New York: Free Press, 1980), 20–62; Michael Massing, *The Fix* (Berkeley: University of California Press, 1998); Solomon H. Snyder, *Brainstorming: The Science and Politics of Opiate Research* (Cambridge, Mass.: Harvard University Press, 1989).

7. Goldberg, "Federal Government's Response," 58–59.

8. Cassandra Tate, *Cigarette Wars: The Triumph of "The Little White Slaver"* (New York: Oxford University Press, 1999).

9. On the middle class's proclivity for making others give up drugs it has decided to give up, see David T. Courtwright, "The Rise and Fall and Rise of Cocaine in the United States," in *Consuming Habits: Drugs in History and Anthropology*, ed. Jordan Goodman, Paul E. Lovejoy, and Andrew Sherratt (New York: Routledge, 1995), 206–28.

10. Woodruff D. Smith, "From Coffeehouse to Parlour: The Consumption of Coffee, Tea and Sugar in North-Western Europe in the Seventeenth and Eighteenth Centuries," in *Consuming Habits: Drugs in History and Anthropology*, ed. Jordan Goodman, Paul E. Lovejoy, and Andrew Sherratt (New York: Routledge, 1995), 148–64; and, in the same volume, Jordan Goodman, "Excitantia: Or, How Enlightenment Europe Took to Soft Drugs," 126–47.

11. See, e.g., Jon Elster, ed., *Addiction: Entries and Exits* (New York: Russell Sage, 1999).

12. Nathan B. Eddy and Everette L. May, "The Search for a Better Analgesic," *Science* 181 (1973): 407–14; Snyder, *Brainstorming;* Caroline Jean Acker, "Planning and Serendipity in the Search for a Nonaddicting Opiate Analgesic," in *The Inside Story of Medicines: A Symposium*, ed. Gregory J. Higby and Elaine C. Stroud (Madison, Wis.: American Institute for the History of Pharmacy, 1997), 139–57.

13. Allan M. Brandt, *No Magic Bullet: A Social History of Venereal Disease in the United States since 1880* (1985; rev. ed., New York: Oxford University Press, 1987), 122.

14. Ralph Chester Williams, *The United States Public Health Service, 1798–1950* (Washington, D.C.: USPHS, 1951), 388–89.

15. Elizabeth Fee, "Sin Versus Science: Venereal Disease in Twentieth-Century Baltimore," in *AIDS: The Burdens of History,* ed. id. and Daniel M. Fox (Baltimore: Johns Hopkins University Press, 1988), 121–46; James H. Jones, *Bad Blood: The Tuskegee Syphilis Experiment* (1981; rev. ed., New York: Free Press, 1993).

16. Brandt, *No Magic Bullet;* Williams, *Public Health Service,* 783.

17. William L. White, *Slaying the Dragon: The History of Addiction Treatment and Recovery in America* (Bloomington, Ill.: Chestnut Hill Systems, 1998).

18. Jim Baumohl, "Maintaining Orthodoxy: The Depression-Era Struggle over Morphine Maintenance in California," in *Altering American Consciousness: Alcohol and Drugs in American History,* ed. Sarah Tracy and Caroline Jean Acker (Amherst: University of Massachusetts Press, forthcoming).

19. Guenter B. Risse, "Epidemics and History: Ecological Perspectives and Social Responses," in *AIDS: The Burdens of History,* ed. Elizabeth Fee and Daniel M. Fox (Berkeley: University of California Press, 1988), 33–66.

20. Brandt, *No Magic Bullet,* 174–76.

21. Thomas J. Sugrue, "The Structures of Urban Poverty: The Reorganization of Space and Work in Three Periods of American History," in *The "Underclass" Debate: Views from History,* ed. Michael B. Katz (Princeton: Princeton University Press, 1993), 85–177.

22. Don Des Jarlais et al., "Maintaining Low HIV Seroprevalence," *Journal of the American Medical Association* 274 (Oct. 18, 1995): 1226–30.

Acknowledgments

It is humbling to reflect on how many people, in how many ways, have contributed to me as I have worked for over ten years on the project which culminates in the publication of this book. I can't possibly name them all here, but I gladly acknowledge the debt I owe to colleagues, fellow scholars, students, friends, and family. I also recognize that however much help I have received from them, the shortcomings of this book remain entirely my own responsibility.

Gerald Grob and John Parascandola read the entire manuscript; they contributed useful commentary and saved me from several errors. Elizabeth Fee provided a very helpful reading of an early version of Chapter 6. I am grateful to several anonymous reviewers whose suggestions sharpened my thinking and led to substantial revision of several chapters. Among my many supportive colleagues in the History Department at Carnegie Mellon University, Mary Lindemann and David Hounshell have been unstinting in providing feedback as the manuscript evolved through several stages. Thanks to Peter Dreyer for his deft and thorough editing and to Jacqueline Wehmueller for her support and wisdom.

Guenter B. Risse, chair of the Department of the History of Health Sciences at the University of California, San Francisco, where I earned my doctoral degree, introduced me to the history of medicine as only a learned scholar and gifted teacher can. Adele E. Clarke was a mentor in many ways: she included me in discussion groups on the sociology of knowledge; she provided meticulous feedback on written work and a steady stream of provocative leads; and, by example and through advice, she helped guide a neophyte scholar in the early stages of a career. I am also grateful to Adele for introducing me to Howard Becker. Meetings of the History of Medicine and Culture Group, sponsored jointly by Guenter Risse and Thomas Laqueur of the University of California, Berkeley, History Department, provided a setting where heated intellectual debate did not prevent warm and enduring friendships from forming. I continue to be grateful for the friendships I formed with David Barnes, Elizabeth Haiken, Robert Martensen, Tina Stevens, and many others. I was fortunate to be at UCSF during the time that Dorothy Porter spent a year there; as a teacher and as a colleague, she set an inspiring example of intellectual excellence combined with personal verve. Conversations with Ron Roizen introduced me to the field of alcohol studies. I thank Tom Laqueur for suggesting to me that history of medicine was the right field for my interests in graduate study.

It saddens me deeply that my graduate advisor, Jack Pressman, did not live to see this book published. His keen intellect and his knowledge of the history of psychiatry and the medical sciences in America exerted a profound influence on the development of this project.

In 1997, Sarah Tracy organized a conference on the history of drugs and alcohol in America sponsored by the Philadelphia College of Physicians. Remarkably, this exciting conference brought together historians and sociologists of drugs and alcohol for the first time. Since then, Sarah's and my ongoing collaboration has been both professionally and personally rewarding.

For almost two decades, David Musto's *The American Disease* and David Courtwright's *Dark Paradise* stood almost alone as serious historical studies of the history of drug laws and drug use in America. I know both Davids have been delighted to witness and support the emergence of a new cadre of historians interested in drug policy, patterns of drug use, and the social and cultural meanings of drugs in America. I am grateful to both of them for their interest in my work and for their deeply knowledgeable criticisms and suggestions.

I am deeply grateful to the groups whose financial support helped make my research possible. While a graduate student at the University of California, San Francisco, I received support in the form of University of California Regents Fellowships, the University of California Presidential Fellowship, and grants from the Graduate Division of the University of California, San Francisco and the Diane Linkletter Fund, Department of History of Health Sciences, University of California, San Francisco. Travel to archives was supported by grants from the Rockefeller Archive Center, the American Institute for the History of Pharmacy, the Wood Institute for the History of Medicine at the College of Physicians of Philadelphia. I also received a Faculty Development Grant from the College of Humanities and Social Sciences, Carnegie Mellon University.

In 1993 and 1994, I was extremely fortunate to be able to spend a year in the Historical Office of the National Institutes of Health as the DeWitt Stetten, Jr., Fellow in the History of Twentieth-Century Biomedical Sciences and/or Technology. This fellowship provided me an unparalleled opportunity to spend countless hours in the Historical Division of the National Library of Medicine, to attend scientific presentations on a wide range of topics related to my research interests, and to spend time with scientists at work in the laboratories, talking and listening. Kenner Rice made available the laboratory notebooks of Lyndon F. Small. Conversations with him and Arthur Jacobson gave me invaluable insight into how the field of opioid pharmacology has developed. Conversations with Everette

May and Louis Harris of the Medical College of Virginia in Richmond deepened my understanding of the relationship of chemistry to pharmacology. Thanks to John Parascandola, formerly Chief of the Historical Division at the National Library of Medicine and now the Historian of the U.S. Public Health Service; from my earliest days as a graduate student, he has helped me understand the history of pharmacology and of the Public Health Service. Thanks also to the community of historians of medicine in the Washington area, especially John Swann and Suzanne White, historians of the Food and Drug Administration. Above all, Victoria A. Harden, director of the NIH Historical Office, provided an intellectual home that gave me crucial time to deepen my thinking about scientific research before I was swept up in teaching duties and the relentless rhythm of semesters. I will always be grateful for her generosity of spirit which manifested itself in boundless logistical, intellectual, and moral support.

Librarians and archivists have guided me to and through manuscript collections with skill and courtesy. They include Aloha South at the National Archives and Records Administration; the staff of Historical Division of the National Library of Medicine; and Janice Goldblum of the National Academy of Sciences/National Research Council Archives. Michael Aldrich made available works from his astounding collection of works on drugs. Thomas Rosenbaum of the Rockefeller Archive Center brought to my attention numerous sources I would not otherwise have found. I also thank the staff of the Archives of the University of Pennsylvania where the Alfred Newton Richards Papers are held. At the College of Physicians of Philadelphia, Thomas A. Horrocks (now at the Countway Library at Harvard University), Charles B. Greifenstein, Curator of Archives and Manuscripts, and Monique Bourque made repeated visits pleasurable and productive. Charles Rosenberg made me aware that the College held the Records of the Philadelphia Committee for Clinical Study of Opium Addiction; in doing so, he opened up a vista that took my research and thinking in new directions and enabled me to add material on the perspectives and lives of opiate addicts in the first decades of the twentieth century.

During extended stays in Philadelphia, Jennifer Gunn, Janet Golden and Eric Schneider, Judy Porter, Janet Tighe, and Henrika Kuklick provided hospitality and intellectual stimulation. Sue and Maitland Sharpe and Merrie and Ricardo Blocker-Merlo housed me during research visits to Washington, D.C.

Years ago, I began working in the drug field as director of Up Front Drug Information, a community based drug information and education center in Miami. Up Front was founded by activists convinced that accurate information about drugs was better than scare tactics and that drug users had a right to the knowledge and technological means to reduce the risk of harm associated with drugs. In this setting, I learned the basic principles of harm reduction years before that phrase had come to label a movement. Tracy Brown, Alice Brown, and Patricia (Cleckner) Morningstar all contributed to my education there. David E. Smith and the staff of the Haight Ashbury Free Clinic became valued colleagues then and have continued to be so as I shifted from front line work to academic study in the field of drugs. Jim Dubé's friendship has been an enduring source of support and learning.

I returned to harm reduction in 1991 when I became involved in needle exchange work, first in Marin County, California, and then in Pittsburgh, Pennsylvania. This work has informed my research in ways I never anticipated. Following the policy debates between harm reductionists and advocates of zero tolerance has made me aware of the continuing negative impact of drug laws conceived in the Progressive Era on the lives and health of injection drug users, their families, and their communities—our community. Interacting over the course of years with injection drug users who have been partners in the work of public health outreach helped me interpret the experiences of morphine and heroin addicts in the 1920s as I read the transcripts of interviews with them. David True was my first public health outreach teacher; he has remained an inspiration to me ever since. In Pittsburgh, James Crow had paved the way for founding a street-based needle exchange program; as we worked together to found Prevention Point Pittsburgh in 1995, I benefited from his knowledge of this city, the issues surrounding harm reduction, and the nature of activism. I continue to benefit from his love and friendship. Stuart Fisk and Alice Bell have become friends and partners on whom I rely in every possible way. Kellie Jones, Donna Riley, Lisa Sigel, Carol Moeller, Wesley Light, Laura Horowitz, Nancy Bernstein, Michael Grosberg, Amy Cyphert, Linda Ogden, Mack Friedman, Lisa Ellinger Gorman, and the other volunteers of Prevention Point Pittsburgh, past and present, have sustained an ongoing laboratory on how, in Margaret Mead's words, "a small group of thoughtful committed citizens can change the world."

Words are inadequate to recognize the combination of intellectual and personal support provided by my colleagues Judith Modell, Mary Lindemann, and Scott Sandage. Sharon Ghamari-Tabrizi reminded me to steer clear of the shoals of pharmacological reductionism when she said, "I can't imagine thinking about drugs without thinking about consciousness." John Hinshaw's loving friendship has gladdened happy times and buoyed me through discouragement. His careful reading of several chapters and ongoing conversations have helped me understand what it is to be an activist scholar.

My children, Natasha and John Ciancutti, responded to their mother's quixotic decision to return to graduate school in incipient middle age with wit, grace, and loving support—just one small reason why they remain the central source of happiness in my life. My brother Peter Acker has always been an inspiration; I am grateful for the continuity of his love and support even though demanding lives in different cities mean we don't spend nearly as much time together as I would like. His work as a pediatrician exemplifies what it means to be a wise and humane physician. Vera Acker, who married my father in 1975, brought love and grace to a grieving family; she and my father, Robert Acker, have created a home that has been a haven for me at semesters' end and whenever I have needed rest and restoration. My father's love of reading and his pride in my early successes in school helped set the course which led to my becoming a scholar. Throughout my graduate schooling, his material and moral support were critical to my continuing on to finish my degree. I only hope this book gives him another reason to feel proud of his daughter.

Index